"中国森林生物多样性监测网络"丛书 马克平 主编

广西弄岗喀斯特季节性雨林
——树种及其分布格局

Guangxi Nonggang Karst Seasonal Rain Forest: Tree Species and Their Distribution Patterns

王 斌 黄俞淞 李先琨 向悟生 丁 涛 刘晟源
刘 演 陆树华 农重刚 陆茂新 韩文衡 李冬兴 著

中国林业出版社
China Forestry Publishing House

图书在版编目（CIP）数据

广西弄岗喀斯特季节性雨林——树种及其分布格局/王斌等著.—北京：中国林业出版社，2016.11
（中国森林生物多样性监测网络丛书）

ISBN 978-7-5038-8608-9

I.①广… II.②王… III.①喀斯特－季节性雨林－树种－研究－广西 IV.①S718.54

中国版本图书馆CIP数据核字(2016)第155189号

内容简介

本书详细介绍了广西西南部原生性喀斯特季节性雨林的217个树种，对每个树种进行形态特征描述，并配备3张精美照片，展示植物的树干、小枝、花序、果实或幼苗等；同时附有每个树种在广西弄岗喀斯特季节性雨林15 hm^2样地内的个体数量、重要值排序、径级分布表及空间分布图，介绍植物种群的优势度、径级结构及生境偏好等。阐述弄岗喀斯特季节性雨林区域的地形地貌、植被特征、生物多样性及其特有性，并对弄岗森林样地内的生境特征、土壤状况与植物群落结构等详细说明。本书资料翔实、图片精美，是了解和研究北热带喀斯特季节性雨林不可多得的参考书，也可作为相关专业研究人员和植物学爱好者了解喀斯特森林、认识森林植物的野外指导手册。

中国林业出版社·科技出版分社

策划、责任编辑：于界芬　于晓文
特约编辑：李莉
电话：010-83143542

出　版：中国林业出版社（100009　北京西城区德内大街刘海胡同7号）
网　址：http://lycb.forestry.gov.cn/
发　行：中国林业出版社
印　刷：北京卡乐富印刷有限公司
版　次：2016年11月第1版
印　次：2016年11月第1次
开　本：889mm × 1194mm　1 / 16
印　张：16
字　数：403千字
定　价：168.00元

序 言 1

在过去的几十年时间里，中国科学院和林业、农业等相关部门陆续建立了数百个生态系统定位研究站。其中，中国科学院组建的中国生态系统研究网络 (CERN) 拥有分布于全国包括农田、森林、草地、湿地、荒漠等生态系统类型的36个生态站。国家林业局建立的中国森林生态系统研究网络 (CFERN) 由29个生态站组成，基本覆盖了我国典型的地带性森林生态系统类型和最主要的次生林、人工林类型。

随着研究的发展，特别是近年来人们对生物多样性和全球变化研究的关注，国际上正在推动生态系统综合研究网络平台的建立。在全球水平上，全球生物多样性综合观测网络 (GEO BON) 是一个有代表性的研究网络。它试图把全球与生态系统和生物多样性长期定位研究相关的网络整合起来，通过综合研究，探讨生态系统与生物多样性维持与变化机制以及系统之间的相互作用机理，为生态系统可持续管理与生物多样性的保护提供科学依据和管理模式。

近年来，中国科学院生物多样性委员会组织建立了中国森林生物多样性监测网络（Chinese Forest Biodiversity Monitoring Network，以下简称CForBio）。中国是生物多样性特别丰富的少数国家之一，也是唯一一个具有从北部寒温带到南部热带完整气候带谱的国家。截止2016年，中国森林生物多样性监测网络包括大型监测样地18个，成为继美国史密森研究院热带研究所建立的热带森林生物多样性监测网络（CTFS-Forest GEO）之后又一大型区域监测网络。由于CForBio横跨多个纬度梯度，对于揭示中国森林生物多样性形成和维持机制，以及森林生物多样性对全球变化的响应，科学利用和有效保护中国森林生物多样性资源具有重要意义。

目前，CForBio已经有很好的研究进展，各样地研究成果陆续在国际著名生态学刊物如*Ecology, Journal of Ecology, Oikos*等上发表，受到国内外同行的高度评价。但这些文章都是关于某一具体问题的研究总结，还无法让国内外同行全面了解CForBio各个样地整体情况。因此，出版这套以中英文形式介绍各大样地基本情况的“中国森林生物多样性监测网络”丛书是非常必要的。感谢马克平研究员组织相关专家编写这套丛书。我相信该丛书不仅是国内外同行深入了解CForBio各样地的参考书，同时也将为我国森林生物多样性监测和森林生态系统联网研究奠定重要的基础。

（孙鸿烈）

中国科学院前副院长

Foreword 1

In the past few decades, hundreds of Ecosystem Research Stations have been set up by the Chinese Academy of Sciences, State Forestry Administration, Ministry of Agriculture and other relative departments. Among them, 36 ecological research stations were established by Chinese Ecosystem Research Network (CERN), supported by the Chinese Academy of Sciences. The 36 research stations are scattered over the country representing diverse ecosystems, including farmland, forest, grassland, wetland, desert and others. Moreover, the Chinese National Ecological Research Network (CFERN), supported by the State Forestry Administration, consists of 29 research stations, covering typical zonal forest ecosystems and main secondary forests and plantations in China.

With the development of research, especially the growing concern over researches on biodiversity and global change in recent years, the establishment of ecosystem research network have been promoted under international supports. So the Group on Earth Observations Biodiversity Observation Network (GEO-BON) is representative across the world, and it attempts to integrate worldwide networks relating to long-term research on ecosystem and biodiversity. Based on the comprehensive studies, the maintenance and change mechanism of ecosystem and biodiversity and their interactions have been explored, which provide scientific basis and management mode for sustainable development of ecosystem and protection of biodiversity.

In recent years, Chinese Forest Biodiversity Monitoring Network (CForBio) has been built by Biodiversity Committee of the Chinese Academy of Sciences. China is one of the few top "mega-biodiversity countries" in the world, and it is also the only country with full climatic zone spectrum, ranging from northern cool temperate zone to southern tropical zone. By the end of 2016, 18 forest dynamics plots have been set up for CForBio, being another large regional monitoring network after the global forest biodiversity monitoring network (CTFS-Forest

GEO). As being across several latitudinal gradients, CForBio is of great significance to reveal the formation and maintenance mechanism of forest biodiversity in China, their response to climate change and their scientific use and effective conservation.

Encouraging progress has been made in this area since the network built, for lots of research findings have been published in the international peer reviewed ecological journals, such as *Ecology, Journal of Ecology* and *Oikos,* etc., which brought about positive response from colleagues in the field of plant ecology. However, the published papers mostly focus on research of specific problems; scientists and public still can't understand the whole situation of each plot in details. So it is really necessary to publish this series, which introduce basic information of permanent forest plots in both Chinese and English. I am grateful to Professor Keping Ma for organizing related specialists to prepare the series. And I believe that this series would be a valuable reference book for scientists and public to further understand CForBio, and it will also lay a foundation for the forest biodiversity monitoring and forest ecosystem research in China.

Honglie Sun
The former Vice-President for the Chinese Academy of Sciences

序 言 2

森林在维持世界气候与水文循环中起着根本性的作用。森林是极为丰富多样的动物、植物与微生物的家园，而人类正是依靠这些生物获取各种产品，包括食品与药物。尽管对人类福祉如此重要，森林仍然遭受着来自土地利用与全球气候环境变化的巨大威胁。在这种不断变化的情况下，为了更好地管理全球剩余的森林，迫切需要树种在生长、死亡与更新方面的详细信息。

中国森林生物多样性监测网络 (CForBio) 正在中国沿着纬度与环境梯度建立大尺度森林监测样地。通过这个重要的全国行动倡议与来自中国科学院及若干其他单位的研究者的努力，CForBio开始搜集关于中国森林的结构与动态的关键信息。现在CForBio与史密森研究院及哈佛大学阿诺德树木园的热带森林监测网络 (CTFS) 形成了合作伙伴。CTFS是个在24个热带或温带国家拥有长期大尺度森林动态研究样地的全球性网络。CForBio与CTFS合作的目标是通过合作研究，了解森林是如何运作的，它们是如何随着时间而改变的，以及如何重建或者恢复，以确保森林提供的环境服务能可持续或者增长。森林及其提供的服务的长期可持续性有赖于我们预测森林对全球变化，包括气候与土地利用变化的响应的能力，以及我们去理解与创建适当的森林服务市场的能力。通过拥有63个森林大样地的全球网络及大量项目的训练与能力建设，CForBio与CTFS的伙伴关系是发展这些预测工具的重要基础。这种伙伴关系也将促进为全球各地的当地社区、林业管理者与政策制定者在森林的保育与管理方面发展应用性的林业项目建议，发展与示范利用乡土物种进行森林重建的方法，以及从经济学角度评估森林在减缓气候变化、生物多样性保护和流域保护上的价值的方法。

我祝贺作者们创作了这部关于样地植物的优秀丛书。本丛书为将来的森林监测提供了基准信息，是涉及森林恢复、碳存储、动植物关系、遗传多样性、气候变化、局地与区域保育等研究内容的研究者、学生与森林管理者们有价值的参考资料。

S. J. 戴维斯
主任
史密森热带研究所＆哈佛大学阿诺德树木园
热带森林科学研究中心

Foreword 2

Forests play an essential role in regulating of world's climatic and hydrological cycles. They are home to a vast array of animal, plant and microorganism species on which humans depend for many products, including food and medicines. Despite the importance of forests to human welfare they are under enormous threat from changes in land-use and global climatic conditions. In order to better manage the world's remaining forests under these changing conditions detailed information on the dynamics of growth, mortality and recruitment of tree species is urgently needed.

The Chinese Forest Biodiversity Monitoring Network (CForBio) that aims to establish large-scale forest monitoring plots across latitudinal and environmental gradients in China. Through this important national initiative, researchers from the Chinese Academy of Sciences and several other research institutions in China, CForBio has begun to gather key information on the structure and dynamics of China's forests. The CForBio initiative is now partnering with the Center for Tropical Forest Science (CTFS) of the Smithsonian Research Institute and the Arnold Arboretum of Harvard University. CTFS is a global program of long-term large-scale forest dynamics plots in 24 tropical and temperate countries/areas. The goal of the partnership between CForBio and CTFS is to work together to understand how forests work, how they are changing over time, and how they can be re-created or restored to ensure that the environmental services provided by forests are sustained or increased. The long-term sustainability of forests and the services they provide depend on our ability to predict forest responses to global changes, including changes in climate and land-use, and our ability to understand and create appropriate markets for forest services. The CForBio-CTFS partnership is ideally poised to develop these predictive tools through a global network of 63 large forest plots and an extensive program of training and capacity building. The partnership will also lead to the development of applied forestry programs that advise local communities, forest managers and policy makers around the world on conservation and management of forests, to develop and demonstrate methods of native species reforestation, and to economically value the roles that forests play in climate mitigation, biodiversity conservation, and watershed protection.

I congratulate the authors on the production of this excellent new series of stand books. In addition to providing a baseline for future forest monitoring, these books provide a valuable resource for researchers, students, and forest managers dealing with issues of forest restoration, carbon storage, plant-animal interactions, genetic diversity, climate change, and local and regional conservation issues.

Stuart Davies

Director

Center for Tropical Forest Science / Forest GEO

The Smithsonian Tropical Research Institute &

The Arnold Arboretum of Harvard University

前　言

我国西南喀斯特地区面积约54万km^2，是全球喀斯特集中连片分布面积最大、岩溶发育最强烈的地区，也是石漠化最严重和生态环境最脆弱的地区之一。

原始生境下保存完好的喀斯特森林生态系统，在维系区域生态安全方面发挥着重要作用，也可作为石漠化治理及生态修复工程中的参考系统，并提供丰富的种源，具有巨大的生态学、社会学和经济学价值。

北热带喀斯特季节性雨林是在我国热带北缘喀斯特地区分布的典型森林植被类型。由于热带季风气候、峰丛洼地的景观特征、复杂的生境类型及富钙偏碱的地球化学背景等影响，该森林呈现出以热带植物区系为主、群落结构复杂、树种组成多样、特有物种丰富等特点。

广西弄岗国家级自然保护区仍保存着世界少有、面积较大、生态系统完整的典型喀斯特季节性雨林，代表树种有蚬木(*Excentrodendron tonkinense*)、肥牛树(*Cephalomappa sinensis*)、东京桐(*Deutzianthus tonkinensis*)、金丝李(*Garcinia paucinervis*)等，是一座独特而巨大的生物多样性基因宝库，是我国14个具有国际意义的陆地生物多样性关键地区之一。因此，广西弄岗喀斯特季节性雨林的森林生态学研究在学术理论和经济社会影响方面均具有深远的意义。

早在明代，我国地理旅行家徐霞客(1586—1641)就考察并记录了中国西南部的广西、贵州和云南一带的喀斯特地貌的景观特征，这是最早关于喀斯特地貌的记载。

我国对喀斯特森林的调查研究可追溯到20世纪三四十年代，当时有少数科学家（如苏宏汉1935，钟济新和陈立卿1940等）在该区域采集植物标本，同时描述植被状况。到20世纪50年代，我国热带、亚热带生物资源考察工作大规模开展，包括对部分喀斯特森林的考察，如1952—1955年华南热作垦殖勘察，1958—1961年华南热带生物资源综合考察，1957—1961年云南热带、亚热带生物资源综合考察等。

我国对喀斯特季节性雨林的大规模考察始于1979年，广西植物研究所主持了由12个专业组成的队伍开展广西弄岗自然保护区综合考察，包括了对季节性雨林地质地貌、植物区系、植被类型等的详细调查和记录。

这些调查研究积累了关于喀斯特森林物种组成、植物区系和植被类型等方面的丰富资料。然而，这些考察仍属于以定性调查为主的研究方式。

20世纪80年代数量生态学和种群生态学等研究方法不断引入，我国森林生态学相关研究得到了迅速发展，陆续出现较多定量研究的工作报道——在喀斯特森林群落组成、结构、分布格局、种间关联、生境关联和生态系统功能等方面，也积累了丰富的资料和经验。

然而，随着生态学学科的发展和深入，以往一些依赖于小面积、短期群落样方的调查数据逐渐无法满足需求。因为很多关键的森林生态学过程（如生境过滤、种子扩散、种群更新等过程）是在不同的空间和时间尺度上发生作用的，想要更加全面准确地揭示这些森林生态学作用过程、阐明森林生物多样性的形成与维持机理，就迫切需要更加规范、系统的研究方法及翔实的长期监测数据。

美国Smithsonian热带研究所的热带森林科学研究中心(Center for Tropical Forest Science，CTFS)和普林斯顿大学于1982年在巴拿马建立了世界上第一个大型固定监测样地（BCI热带雨林50hm^{2}监测样地），随后发起了全球森林生物多样性监测网络(CTFS-ForestGEO)的建设。

截至2015年，该网络包括了建立于全球24个国家和地区的63个大型固定森林监测样地，定位并挂牌监测的木本植物达到1万种600万株，验证和发展了生物多样性形成及维持的诸多理论和假说，在生态学领域产生了巨大的影响，为全球森林监测研究提供了更加规范的研究方法。

中国森林生物多样性监测网络(Chinese Forest Biodiversity Monitoring Network，CForBio)由中国科学院生物多样性委员会于2004年启动建设，是中国森林生物多样性变化的监测基地，也是CTFS-ForestGEO网络的重要组成部分。截至2016年，CForBio森林监测网络包括了18个大型固定监测样地，涵盖了不同纬度带的森林植被类型，包括北方针叶林、针阔混交林、落叶阔叶林、常绿落叶阔叶混交林、常绿阔叶林以及热带雨林等。其中，广西弄岗北热带喀斯特季节性雨林15 hm^{2}监测样地于2011年底建成，是CForBio和CTFS-ForestGEO网络中唯一的喀斯特季节性雨林监测样地。

弄岗喀斯特季节性雨林15 hm^{2}监测样地，为后续长期翔实的科学数据监测和深入的喀斯特森林生态学研究提供良好的野外科学平台，相关研究成果可为生物多样性保育、喀斯特森林生态系统管理、石漠化区域生态建设工作提供科学的理论指导。

本书介绍了喀斯特季节性雨林的地貌、土壤、植被特征等，基于弄岗样地2011年植被和地形数据，展示弄岗样地内的树种组成、径级结构、空间分布格局和形态特性等，为今后深入开展喀斯特季节性雨林研究提供必要的基础资料；同时配有各树种不同器官的照片，可为读者认知植物提供最直接的感性信息。我们期望本书能吸引更多的同行到弄岗喀斯特森林开展合作研究，也希望广大青年学子加入森林生态学的探索行列。

由于时间仓促，水平有限，错误疏漏在所难免，敬请各位读者不吝赐教。

李先琨　王斌

2016年10月

Preface

The karst area of southwest China, with an area of about 540 thousand km^2, is the largest karst area of concentrated and contiguous distribution and the most strongly developed karst area in the world, and is also one of the most serious rocky desertification and ecologically fragile areas.

The primitive well-preserved karst forest ecosystem in the original habitat plays an important role in maintaining regional ecological security, can be used as an important reference system and provides substantial seed source for the ecosystem reconstruction project in the rocky desertification area, and has great ecological, sociological and economical potential value.

The northern tropical karst seasonal rain forest, located on the northern margin of the tropical limestone karst region of China, is the typical forest vegetation type in this region. The karst forest displays various characteristics, such as obvious tropical floristic elements, complicated community structure, high species richness, and abundant endemic biological species, owing to the tropical monsoon climate, landform characteristics of peak cluster depression, heterogeneous habitats, and special limestone soils with abundant Ca and relatively high pH.

Guangxi Nonggang National Nature Reserve, which has well preserved the typical karst seasonal rain forest until now with large area, integrity ecosystem and representative species such as *Excentrodendron tonkinense, Cephalomappa sinensis, Deutzianthus tonkinensis,* and *Garcinia paucinervis*, etc., is an unique and great genetic treasure house of biodiversity, and one of the 14 key areas with international significance in China. Therefore, the forest ecological researches on the karst seasonal rain forest in Nonggang, Guangxi have far-reached significance in both academic theory and economic and social influence.

As early as the Ming Dynasty, China's geographic traveler Xiake Xu (1586-1641) investigated and recorded the karst landscape characteristics in Guangxi Zhuang Autonomous Region, Guizhou and Yunnan provinces in Southwest China. It is the earliest record about the karst landscape.

Researches on karst forest vegetation can be traced back to the 1930s and 1940s, at that time a few scientists (such as, Honghan Su 1935, Jixin Zhong and Liqing Chen 1940, et al.) described the vegetation distribution of karst forest when they collected plant specimens in the region.

In the 1950s, the tropical and subtropical biological resources investigations were initiated on a large scale in China, including some investigations of karst forests in South China, such as the tropical cultivation survey in South China from 1952 to 1955, the comprehensive survey of the tropical biological resources in South China from 1958 to 1961, and the comprehensive survey of the tropical and subtropical biological resources in Yunnan province from 1957 to 1961, and so on .

The large-scale investigation of karst seasonal rain forest began in 1979. Guangxi Institute of Botany presided over the comprehensive survey of Guangxi Nonggang National Nature Reserve, with a survey team consisting of 12 professional groups. In this survey, the flora and vegetation types of the northern tropical karst seasonal rain forest were investigated.

These surveys collected much information on the species composition, flora, and vegetation types of karst forests. However, these surveys and researches were descriptive, and this period may be considered as a qualitative phase in the forest studies in China.

Following a number of introductions and translations of foreign references on quantitative ecology and population ecology in the 1980s, the research of forest ecology developed rapidly in China, and many quantitative researches were conducted. Abundant information and experiences have been accumulated focusing on the community composition, structure, distribution pattern, interspecific association, habitat association and ecological functions of karst forests.

However, with the development and deepening of ecology, the data from short term surveys and small size plots gradually can not satisfy the demand of ecological research. It is because many of the key forest ecological processes (such as habitat filtering, seed dispersal, and population regeneration process, etc.) worked at different spatial and temporal scales, and if we want to understand these forest ecological processes and the mechanisms for the formation and maintenance of biodiversity more comprehensively and accurately, more standardized system of research methods and complete long-term detailed forest monitoring data are urgently needed .

The Center for Tropical Forest Science (CTFS) of the Smithsonian Research Institute and the Princeton University established the world's first large-scale forest plot (BCI 50km^2 plot) in Panama in 1982, then launched the global forest biodiversity monitoring network (CTFS-ForestGEO).

By 2015, the network included 63 large forest plots in 24 countries and regions in the world, mapping and tagging woody plants reached 6 million individuals in 10 thousand species, verified and developed many theories and hypotheses of biodiversity formation and maintenance, produced huge influence in the ecological field, and provided more standardized research methods for the global forest monitoring studies.

Chinese Forest Biodiversity Monitoring Network (CForBio), which started to build by Biodiversity Committee, the Chinese Academy of Sciences in 2004, is a monitoring base for the change of forest biodiversity in China, is also an important part of the CTFS-ForestGEO network. By 2016, CForBio network included 18 large permanent monitoring plots, covering forest vegetation types at different latitudinal zones, including boreal forest, mixed coniferous and broad leaved forest, deciduous broad-leaved forest, evergreen deciduous broad-leaved mixed forest, evergreen broad-leaved forest, and tropical rain forest . Among them, Nonggang karst seasonal rain forest 15 hm^2 plot, which was completed by the end of 2011, is the unique karst seasonal rain forest plot in CForBio and CTFS-ForestGEO network .

The plot of 15 hm^2 in Nonggang karst seasonal rain forest will provide a powerful platform for long-term scientific monitoring and further ecological study of the karst forest, and the results will provide more sound theoretical guidance for the ecological restoration work in the rocky desertification karst region.

This book introduces the landscape, soil, and vegetation characteristics of karst seasonal rain forest, concretely describes the tree species composition, diameter class structure, spatial distribution pattern, and biological characteristics in Nonggang plot based on the vegetation and terrain data collected in 2011, and these contents were basic information for the future in-depth ecological study. This book also has a lot of exquisite plant photos, which can provide the most direct perceptual materials for readers to acquaint the karst seasonal rain forest and plants. We hope that this book will attract more colleagues to Nonggang karst forest to carry out cooperative researches and more young students to join the research team in forest ecology.

Due to the limitation of time and our knowledge, there might be unavoidable mistakes in this book, and we welcome all comments and suggestions from our readers.

Xiankun Li and Bin Wang

October, 2016

目　录

Contents

广西弄岗国家级自然保护区简介
Brief Introduction to Guangxi Nonggang National Nature Reserve

I

1.1 发展简史

广西弄岗国家级自然保护区前身是弄岗林区和陇瑞林区。其中，弄岗林区于 1978 年被区划界定，1979 年 5 月经广西壮族自治区人民政府批准成立弄岗自治区级自然保护区，1980 年经国务院批准为弄岗国家级自然保护区；陇瑞林区于 1979 年被区划界定，1982 年经批准成立陇瑞自治区级自然保护区。1994 年 9 月经广西区人民政府批准，弄岗国家级自然保护区与陇瑞自治区级自然保护区合并，成立广西弄岗国家级自然保护区。广西弄岗国家级自然保护区于 1999 年加入中国“人与生物圈”保护区网络。

1.1 Brief History

Guangxi Nonggang National Nature Reserve, formely known as the Nonggang Forest Farm and the Longrui Forest Farm. Nonggang Forest Farm was zoned and defined in 1978, approved as Nonggang Provincial Nature Reserve by Guangxi Zhuang Autonomous Region Government in May 1979, and as Nonggang National Nature Reserve by State Council in 1980; Longrui Forest Farm was zoned and defined in 1979, approved as Longrui Provincial Nature Reserve in 1982. Approved by Guangxi Zhuang Autonomous Region Government in September 1994, Guangxi Nonggang National Nature Reserve was established by incorporating Longrui Provincial Nature Reserve with Nonggang National Nature Reserve. Guangxi Nonggang National Nature Reserve joined the Chinese “Man and the Biosphere Programme” Reserves Network in 1999.

1.2 地理位置和自然环境

广西弄岗国家级自然保护区位于广西壮族自治区龙州县之东及宁明县以北（106°42′28″~107°4′54″ E，22°13′56″~22°39′9″ N），总面积 10,077.5 hm^2（图 1）。本区属北热带季风气候，年均气温 22°C，最冷月份平均气温在 13°C 以上，每年有 7 个月的月平均温度在 22°C 以上，极端最高温 40.5°C，极端最低温 −3°C，≥10°C 的年积温 7,834°C，年平均降雨量 1,150~1,550 mm，最多达 2,043 mm，最少 890 mm，其中 76% 雨量集中于 5~9 月（王斌等，2014）。

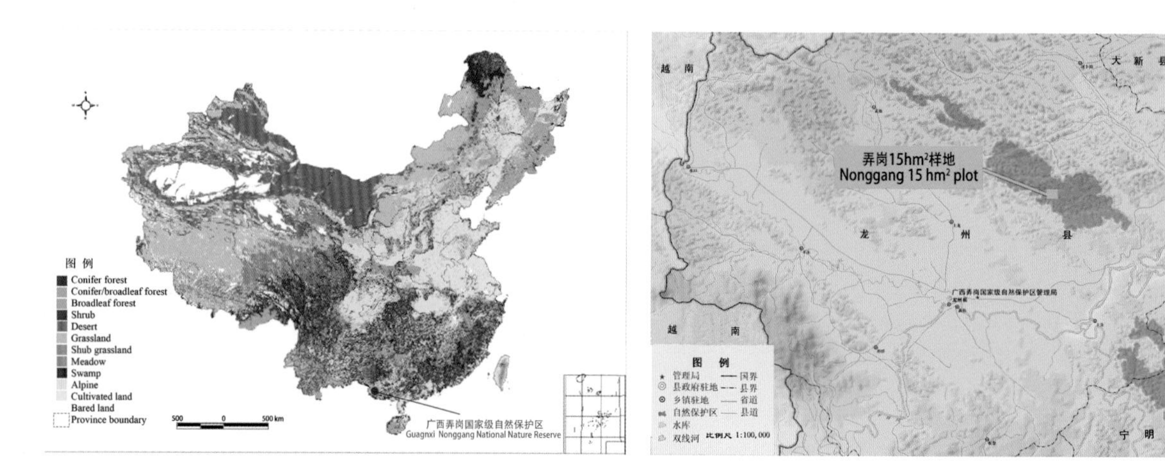

图 1 广西弄岗国家级自然保护区及弄岗 15 hm^2 森林样地的位置

Figure 1 The location of Guangxi Nonggang National Nature Reserve and the 15 hm^2 Nonggang forest plot

保护区为北热带裸露型喀斯特区域，地貌以峰丛深切圆洼地为主，该地貌由多个石山山峰和镶嵌其中的洼地（谷地）组成（图 2）。山峰顶部海拔 400~500 m 左右，最高峰海拔 680.1 m，山峰平均密度 10~30 个/km^2；洼地底部海拔 150~200 m 左右，最低海拔 118.2 m，洼地最大深度 114 m，最大宽度 450 m（李克因，1988）。该区域地表水系缺乏，以暂时性洼地壅水和少量季节性小溪为主；地下水通道丰富，形成复杂的地下河系，枯水期（12 月 ~ 翌年 2 月）最低水位埋深 5~25 m，丰水期（5~7 月）最高水位高出地面 0~3 m（胡长庚，1988）。

根据弄岗喀斯特地貌水分及土壤条件，该喀斯特"峰丛—洼地"坡面的生境可分 3 种类型。①山顶周围：基岩裸露达 95%，全天阳光直射，夏季地表最高温达 60°C 以上，黑色石灰土分布于岩隙间，土层厚度 30~50 cm；②山坡中部：基岩裸露 80% 左右，光照充足，湿度中等，棕色石灰土以斑块状分布，土层厚度 30~60 cm；③洼地周围：岩石裸露 10% 左右，受周围山体遮挡，荫蔽性高，有被水淹的可能，水化棕色石灰土连片分布，土层深厚达 60 cm 以上（苏宗明等，1988）。

1.2 Location and Natural Environment

Guangxi Nonggang National Nature Reserve, approximately 10,077.5 hm^2, is located in east of Longzhou County and north of Ningming County of Guangxi Zhuang Autonomous Region (106°42′28″–107°4′54″ E, 22°13′56″–22°39′9″ N) (Figure 1). The region has a tropical monsoon climate. Mean annual temperature in the region is 22°C, mean temperature of the coldest month is above 13°C, seven months of each year have monthly mean temperature above 22°C, extreme maximum temperature is 40.5°C, extreme minimum temperature is −3°C, and total annual accumulated temperature with mean daily temperature above 10°C is 7,834°C; mean annual precipitation is 1,150–1,550 mm, maximum annual precipitation is 2,043 mm, minimum annual precipitation is 890 mm, and 76% of the rainfall is concentrated between May and September (Wang *et al.*, 2014).

图 2 弄岗自然保护区喀斯特峰丛洼地景观

Figure 2 Karst "Peak-cluster depression" platform in Nonggang Nature Reserve

The area is a northern tropical exposed karst region. The karst geomorphological type is “peak-cluster deeply incised round depression (valley)”, which is composed of a plurality of mountain peaks and depressions (valleys) (Figure 2). The tops of peak clusters are 400–500 m above sea level, the highest peak altitude is 680.1 m, and the average cluster density of mountain peaks is 10–30 per km^2; the bottoms of depression valleys are 150–200 m above sea level, the lowest depression altitude is 118.2 m, the maximum depression depth is 114 m, and maximum depression width is 450 m (Li, 1988). The surface water systems are not well developed in the region, and the surface waters are based on temporary damming waters in depressions and a small amount of seasonal streams. The groundwater channels are abundant, and form a complicated underground river system. The lowest water levels of underground rivers are 5–25 m below ground in low water periods (Dec.–Feb.), and the highest water levels are 0–3 m above ground in high water periods (May–July) (Hu, 1988).

According to the water and soil conditions form the peak to the depression, the “peak-depression” slopes could be divided into three habitats. ①Habitat around peaks: where the rock exposed rate is 95% with direct sunshine all day, the highest temperature is above 60°C in summer, the primitive limestone soils exist in the gaps of rocks, and the soil thicknesses are 30–50 cm. ②Habitat on middle slopes: where the rock exposed rate is 80%, with sufficient sunshine and medium humidity, brown rendzina distribute in patchy pattern, and soil thicknesses are 30–60 cm; ③Habitat around valleys: where the rock exposed rate is 10% with insufficient sunshine because of the occlusion of the surrounding mountains, they might be flooded periodically, hydrated brown rendzina distribute contiguously , and soil thicknesses reach to 60 cm (Su *et al.*, 1988).

1.3 主要植被类型

弄岗保护区的植被类型为北热带喀斯特季节性雨林（植被亚型），保存着丰富的喀斯特特有树种，植物区系以热带性质科属为主（苏宗明等，1988；2014）。与喀斯特“峰丛—洼地”坡面上的三种生境类型相对应，该植被亚型分三个群系组：①铁榄（*Sinosideroxylon pedunculatum*）—清香木（*Pistacia weinmannifolia*）林（群系组），属于旱生型山顶矮林，林高 5~6 m，分布于山顶周围；②蚬木（*Excentrodendron tonkinense*）—肥牛树（*Cephalomappa sinensis*）林（群系组），林高 20~30 m，最大树高达 35 m，分布于山坡中部；③大叶风吹楠（*Horsfieldia kingii*）—望天树（*Parashorea chinensis*）—东京桐（*Deutzianthus tonkinensis*）林（群系组），林高 35 m 左右，最大树高 45 m 以上，板根和茎花现象较普遍，附生及藤本植物较多，具有热带雨林特征，分布于洼地周围（图 3）。

1.3 Main Vegetation Types

The vegetation type of the Nonggang reserve is northern tropical karst seasonal rain forest, preserving abundant karst endemic tree species. The flora, both at the family and genus level, is dominated by tropical floristic elements(Su *et al.*, 1988; 2014). Corresponding to the soil and water conditions in karst peak-cluster depressions, karst seasonal rain forest could be divided into three forest formation-groups. ① *Sinosideroxylon pedunculatum–Pistacia weinmannifolia* forest (Formation-group): it belongs to the xeromorphy elfin forest, the forest canopy height is 5–6 m, and it distributes around peaks. ② *Excentrodendron tonkinense–Cephalomappa sinensis* forest (Formation-group): the canopy height is 20–30 m, the maximum tree height reaches up 35 m, and it distributes in the middle slopes. ③*Horsfieldia kingii–Parashorea chinensis–Deutzianthus tonkinensis* forest (Formation-group): the canopy height is about 35 m, the maximum tree height is up to 45 m, it has the characteristics of tropical rainforest with widespred buttresses and cauliflories, abundant epiphytes and woody lianas, and distributes around valleys(Figure 3).

图 3 弄岗喀斯特季节性雨林植被 Figure 3 Vegetaion of karst seasonal rain forest in Nonggang
A. 喀斯特季节性雨林的整体外貌 The overall appearance of the karst seasonal rain forest;
B. 以毛叶铁榄为优势的山顶矮林外貌 The appearance of the *Sinosideroxylon pedunculatum* var. *pubifolium* community;
C. 以望天树为优势的群落外貌 The appearance of the *Parashorea chinensis* community;
D. 以蚬木为优势的群落内部结构 The internal structure of the *Excentrodendron tonkinense* community;
E. 以肥牛树为优势的群落内部结构 The internal structure of the *Cephalomappa sinensis* community;
F. 以中国无忧花为优势的群落内部结构 The internal structure of the *Saraca dives* community

1.4 物种多样性和特有性

弄岗保护区共计有维管植物 184 科 810 属 1,752 种，其中蕨类植物 29 科 51 属 150 种，裸子植物 4 科 5 属 10 种，双子叶植物 126 科 611 属 1,337 种，单子叶植物 25 科 143 属 255 种（黄俞淞等，2013）。本区植物物种特有性极为明显，蕨类植物中国特有种 39 种、种子植物广西特有种 101 种、喀斯特特有植物 278 种（黄俞淞等，2013）。野生分布的珍稀濒危植物有 33 种：属中国 IUCN 红色名录中极危等级的有水蕨（*Ceratopteris thalictroides*）、望天树和见血封喉（*Antiaris toxicaria*）等 3 种，属中国 IUCN 红色名录中濒危等级的有七指蕨（*Helminthostachys zeylanica*）、五桠果叶木姜子（*Litsea dilleniifolia*）、斜翼（*Plagiopteron suaveolens*）和大叶风吹楠（*Horsfieldia kingii*）等 20 种，属国家一级保护野生植物的有叉叶苏铁（*Cycas bifida*）、石山苏铁（*Cycas spiniformis*）和望天树等 3 种，属国家二级保护野生植物的有蚬木、海南椴（*Diplodiscus trichosperma*）、东京桐、紫荆木（*Madhuca pasquieri*）、黑桫椤（*Alsophilapo dophylla*）和董棕（*Caryota obtusa*）等 16 种。

据已发表的论文及报告统计，弄岗保护区有陆栖脊椎动物 348 种，隶属于 23 目 81 科，其中兽类 7 目 20 科 38 种（嘉道理农场暨植物园，2004；陈天波等，2013），鸟类有 13 目 42 科 222 种（陆舟等，2016），爬行类有 2 目 13 科 61 种（杨岗等，2011），两栖类有 1 目 6 科 27 种（杨岗等，2011）。属国家 I 级保护野生动物有白头叶猴（*Trachypithecus leucocephalus*）、黑叶猴（*Trachypithecus francoisi*）、熊猴（*Macaca assamensis*）、林麝（*Moschus berezovskii*）、蟒（*Python molurus*）、云豹（*Neofelis nebulosa*）、蜂猴（*Nycticebus coucang*）等 7 种；属国家 II 级保护野生动物有猕猴（*Macaca mulatta*）、黑熊（*Selenarctos thibetanus*）、穿山甲（*Manis pentadactyla*）、大灵猫（*Viverra zibetha*）等 36 种。广西特有动物有白头叶猴、弄岗穗鹛（*Stachyris nonggangensis*）、广西林蛇（*Boiga guangxiensis*）、泰北小头蛇（*Oligodn joynsni*）等。其中，2008 年发现的弄岗穗鹛是中国鸟类学家发现、描述并命名的第 2 个鸟类新种。

1.4 Species Diversity and Endemism

In total, 1,752 vascular plant species, belonging to 810 genera and 184 families were recorded in the Nonggang Reserve, including 150 fern species in 51 genera and 29 families, 10 gymnosperm species in 5 genera and 4 families, 1,337 dicotyledon species in 611 genera and 126 families, and 255 monocotyledon species in 143 genera and 25 families (Huang *et al.*, 2013). Among these species, abundant endemic species were recorded, including 39 Chinese endemic fern species, 101 Guangxi endemic seed plant species, and 278 karst endemic species (exclusive calciphytes) (Huang *et al.*, 2013). Thirty-three rare or endangered species were recorded, including 3 critically endangered species of IUCN Red List of China, *Ceratopteris thalictroides*, *Parashorea chinensis*, and *Antiaris toxicaria*, 20 endangered species of IUCN Red List of China, such as *Helminthostachys zeylanica*, *Litsea dilleniifolia*, *Plagiopteron suaveolens*, *Horsfieldia kingii*, etc, 3 species under First-Grade State Protection, *Cycas bifida*, *Cycas spiniformis*, and *Parashorea chinensis*, and 16 species under Second-Grade State Protection, such as *Excentrodendron tonkinense*, *Diplodiscus trichosperma*, *Deutzianthus tonkinensis*, *Madhuca pasquieri*, *Alsophilapo dophylla*, *Caryota obtusa*, etc.

According to statistical results of the published papers and reports, 348 terrestrial vertebrate species, belonging to 23 orders and 81 families, were recorded in the Nonggang Reserve, including 38 mammal species in 7 orders and 20 families (Kadoorie Farm & Botanic Garden, 2004; Chen *et al.*, 2013), 222 bird species in 13 orders and 42 families (Lu *et al.*, 2016), 61 reptile species in 2 orders and 13 families (Yang *et al.*, 2011), and 27 amphibian species in 1 orders and 6 families (Yang *et al.*, 2011). Seven species were under First-Grade State Protection, *Trachypithecus poliocephalus*, *Trachypithecus francoisi*, *Macaca assamensis*, *Moschus berezovskii*, *Python molurus*, *Neofelis nebulosa*, *Nycticebus coucang*; 36 species were under Second-Grade State Protection, such as *Macaca mulatta*,*Selenarctos thibetanus*, *Manis pentadactyla*, *Viverra zibetha*, etc. Four species are Guangxi endemic animal species, *Trachypithecus poliocephalus*, *Stachyris nonggangensis*, *Boiga guangxiensis*, *Oligodn joynsni*, etc. Among them, the *Stachyris nonggangensis* discovered in 2008, is the second new species of birds that were found, described, and named by the Chinese ornithologists.

参考文献:

陈天波，宋亦希，陈辈乐，蒙渊君，温柏豪. 2013. 利用红外线相机监测地表水对广西弄岗国家级自然保护区兽类分布的影响 [J]. 动物学研究，34（3）：145～151.

胡长庚. 1988. 弄岗自然保护区水文地质考察报告 [J]. 广西植物 (增刊 1)：17～32.

黄俞淞，吴望辉，蒋日红，刘晟源，刘演，李先琨. 2013. 广西弄岗国家级自然保护区植物物种多样性初步研究 [J]. 广西植物，33（3）：346～355.

嘉道理农场暨植物园. 2004. 快速生物多样性评估报告—广西西南部弄岗国家级自然保护区 [D]. 中文版第 11 号. 香港：嘉道理农场暨植物园出版.

李克因. 1988. 弄岗自然保护区地貌分区及地貌发育初考 [J]. 广西植物 (增刊 1)：33～51.

陆舟，杨岗，余桂东，赵东东，吴映环，周放. 2016. 广西弄岗国家级自然保护区喀斯特森林鸟类群落结构与多样性分析 [J]. 四川动物，35（1）：141～148.

苏宗明，赵天林，黄庆昌. 1988. 弄岗自然保护区植被调查报告 [J]. 广西植物（增刊 1）：185～214.

苏宗明，李先琨，丁涛，宁世江，陈伟烈，莫新礼. 2014. 广西植被 [D]. 北京：中国林业出版社.

王斌，黄俞淞，李先琨，向悟生，丁涛，黄甫昭，陆树华，韩文衡，文淑均，何兰军. 2014. 弄岗北热带喀斯特季节性雨林 15 ha 监测样地的树种组成及空间分布 [J]. 生物多样性，22（2）：141～156.

杨岗，李东，余辰星，蒋爱伍，蒙渊君，周放. 2011. 广西弄岗国家级自然保护区两栖爬行动物资源调查 [J]. 动物学杂志，46（4）：47～52.

Reference:

Chen TB, Song YX, Chen BL, Meng YJ, Wen BH. 2013. Influence of surface water availability on mammal distributions in Nonggang National Nature Reserve, Guangxi, China [J]. Zoological Research, 34(3): 145–151. (in Chinese with English abstract)

Hu CG. 1988. Report on the investigation of hydrogeology of the Nonggang Nature Reserve [J]. Guihaia (Suppl, 1): 188–214. (in Chinese with English abstract)

Huang YS, Wu WH, Jiang RH, Liu SY, Liu Y, Li XK. 2103. Primary study on species diversity of plants in Nonggang National Nature Reserve of Guangxi [J]. Guihaia, 2013, 33(3): 346–355. (in Chinese with English abstract)

Kadoorie Farm & Botanic Garden. 2004. Report of rapid biodiversity assessments at Nonggang National Nature Reserve, Southwest Guangxi, China, Chinese version No. 11 [D]. Hong Kong: KFBG.

Li KY. 1988. Primary exploration of the geomorphological districts and the development of surface forms in the Nonggang Nature Reserve [J]. Guihaia (Suppl, 1): 33–51. (in Chinese with English abstract)

Lu Z, Yang G, Yu GD, Zhao DD, Wu YH, Zhou F. 2016. Community composition and avian diversity in karst forest of Nonggang National Nature Reserve, Guangxi [J]. Sichuan Journal of Zoology, 35(1): 141–148. (in Chinese with English abstract)

Su ZM, Zhao TL, Huang QC. 1988. The vegetation of Nonggang Nature Reserve in Guangxi [J]. Guihaia (Suppl, 1): 185–214. (in Chinese with English abstract)

Su ZM, Li XK, Ding T, Ning SJ, Chen WL, Mo XL. 2014. The vegetation of Guangxi [D]. Beijing: China Forestry Publishing House. (in Chinese)

Wang B, Huang YS, Li XK, Xiang WS, Ding T, Huang FZ, Lu SH, Han WH, Wen SJ, He LJ. 2014. Species composition and spatial distribution of a 15 ha northern tropical karst seasonal rain forest dynamics plot in Nonggang, Guangxi, southern China [J]. Biodiversity Science, 22(2): 141–156. (in Chinese with English abstract)

Yang G, Li D, Yu CX, Jiang AW, Meng YJ, Zhou F. 2011. Field survey on amphibians and reptiles in Nonggang National Nature Reserve, Guangxi [J]. Chinese Journal of Zoology, 46(4): 47–52. (in Chinese with English abstract)

弄岗喀斯特季节性雨林 $15hm^2$ 监测样地
The 15 hm^2 Karst Seasonal Rain Forest Plot in Nonggang

2

2.1 样地建设和群落调查

弄岗喀斯特季节性雨林 15 hm^2 监测样地，位于广西弄岗国家级自然保护区弄岗片区的弄姆皇（22°25′ N，106°57′ E）。自 2009 年初经过多次实地勘察，使用全站仪测量并调整样地边框，最终确定样地为长方形，东西长 500 m，南北宽 300 m，总面积 15 hm^2。研究团队于 2010 年 7 月至 2011 年 12 月期间完成第一次群落调查。样地建设和群落调查参考 CTFS（Centre for Tropical Forest Sciences）的方法，用全站仪将样地划分成 1,500 个 10 m×10 m 的样方，标定并调查样方内所有胸径（DBH）≥1 cm 的木本植物个体，调查内容包括植物个体的物种名称、胸径、坐标、生长状态等，并挂牌标记。2014 年 10 月在"峰丛—洼地"坡面沿海拔梯度建立土壤监测样带，采集样带内 50 个土壤样方的表层土壤（0~10 cm），测定其理化性质。

2.1 Plot Establishment and Plant Census

The Nonggang Karst Seasonal Rain Forest Dynamics Plot is located in Nongmuhuang, Nonggang National Nature Reserve (22°25′ N, 106°57′ E). After several field surveys in the beginning of 2009, the plot was finally established as a rectangular shape, with size 15 hm^2, 500 m in east-west axis, and 300 m in north-south axis by Total Stations. The first census began in July 2010 and completed in December 2011. Plot establishment and census followed the protocol of CTFS (Centre for Tropical Forest Sciences). With Total Stations, the plot was grided into 1,500 10 m×10 m quadrats. All free-standing trees at least 1 cm in diameter at breast height were tagged, mapped, measured, and identified to species. In October 2014, a soil sample zone was set up along the altitudinal gradient of the "peak-depression" slope in the plot, and the physical and chemical properties of the surface soils (0–10 cm) in 50 soil sample subplots were measured and analyzed.

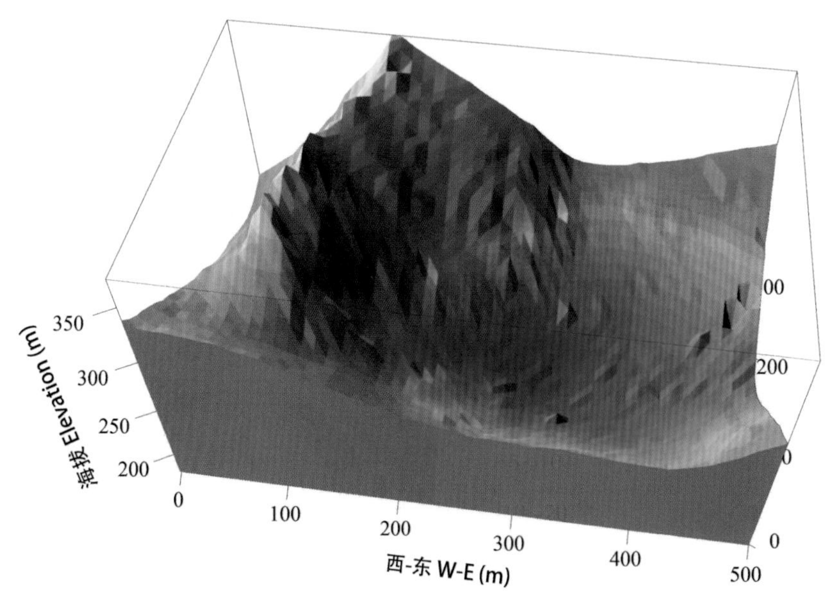

图 4 弄岗样地三维地形图

Figure 4 The three dimension topography map of the Nonggang plot

2.2 地形和土壤

弄岗样地包含了一个小型山峰和一个较完整的洼地，具有从山顶、山坡到洼地等系列完整的“峰丛—洼地”生境类型（图 4）。海拔范围 184~374 m，平均海拔 262.7 m；坡度范围3.7°~78.9°，平均坡度41.7°；山顶周围岩石裸露率 95% 左右，洼地周围岩石裸露率 20% 左右，样地平均岩石裸露率 68.8%（图 5）；绝大部分地段土层厚度不足 30 cm；样地内小生境类型丰富，局部地形复杂多变，部分 10 m×10 m 样方内最大高差达十几米。

样地内土壤理化性质的空间变异性强烈。其中，表层土壤的有机质含量为 58.80~187.60 g/kg，平均 112.09 g/kg；全氮含量为 3.88~11.88 g/kg，平均 7.04 g/kg；全磷含量为 0.53~3.54 g/kg，平均 1.35 g/kg；全硫含量为 2.12~5.7 g/kg，平均 5.74 g/kg；全钾含量为 5.59~11.95 g/kg，平均 8.50 g/kg；镁含量为 1.80~7.70 g/kg，平均 4.31 g/kg；钙含量为 0.30 ~2.90 g/kg，平均 1.21 g/kg；pH 值为 6.30~7.48，平均 6.97。沿“峰丛—洼地”坡面的海拔下降方向，样地表层土壤的有机质、全氮、镁、钙、全硫含量及 pH 值等呈下降趋势，全磷、全钾等含量呈上升趋势（图 6）。

2.2 Topography and Soil

The Nonggang plot includes a small hill and a complete depression, possessing complete karst “peak-depression” habitat types consisting of peak, mountain slope, and depression (Figure 4). The elevation of the plot varies from 184 to 374 m, and a mean of 262.7 m. The slope varies from 3.7° to 78.9°, and a mean of 41.7°. The rock exposed rate is more than 95% around the peak, about 20% around the depression; the mean rock exposed rate of the plot is 68.8% (Figure 5). The soil thickness is less than 30 cm in most of the area. Small habitat types are abundant, and the local terrains are complex and changable. Some elevation differences of 10 m × 10 m subplots are above ten meters.

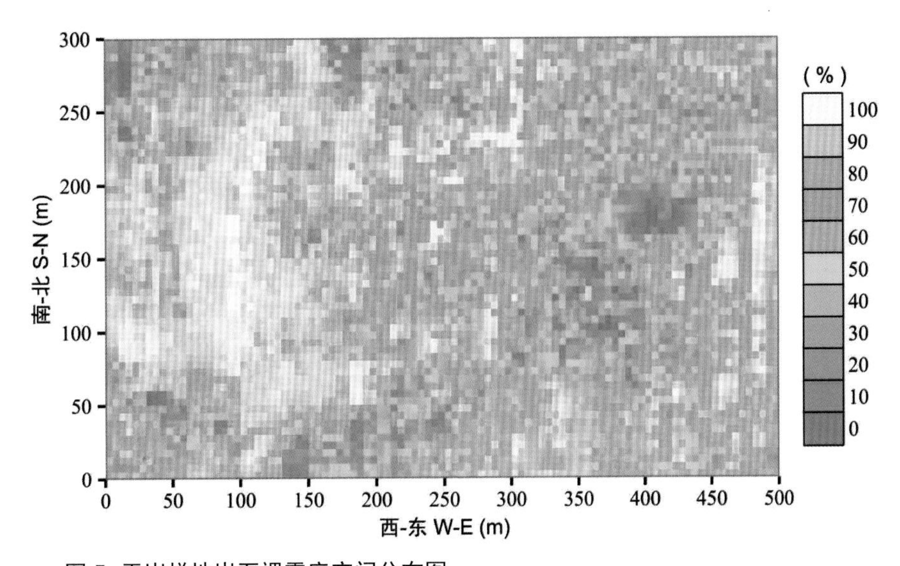

图 5 弄岗样地岩石裸露度空间分布图

Figure 5 Rock exposed degree distribution map of the Nonggang plot

The physical and chemical properties of the surface soils show strong spatial heterogeneity in the soil sample zone. Organic matter content range in 5.88–187.60 g/kg, averaging 112.09 g/kg; total nitrogen (N) content range in 3.88–11.88 g/kg, averaging 7.04 g/kg; total phosphorus (P) content range in 0.53–3.54 g/kg, averaging 1.35 g/kg; total sulfur (S) content range in 2.12–5.7 g/kg, averaging 5.74 g/kg; total potassium (K) content range in 5.59–11.95 g/kg, averaging 8.50 g/kg; magnesium (Mg) content range in 1.80–7.70 g/kg, averaging 4.31; calcium (Ca) content range in 0.30–2.90 g/kg, averaging 1.21 g/kg; pH value range in 6.30–7.48, averaging 6.97. As the altitude of "peak-depression" slope location decreases, the organic matter, total nitrogen, magnesium, calcium, total sulfur contents, and pH value show decreasing trends, total phosphorus and total potassium contents show increasing trends (Figure 6).

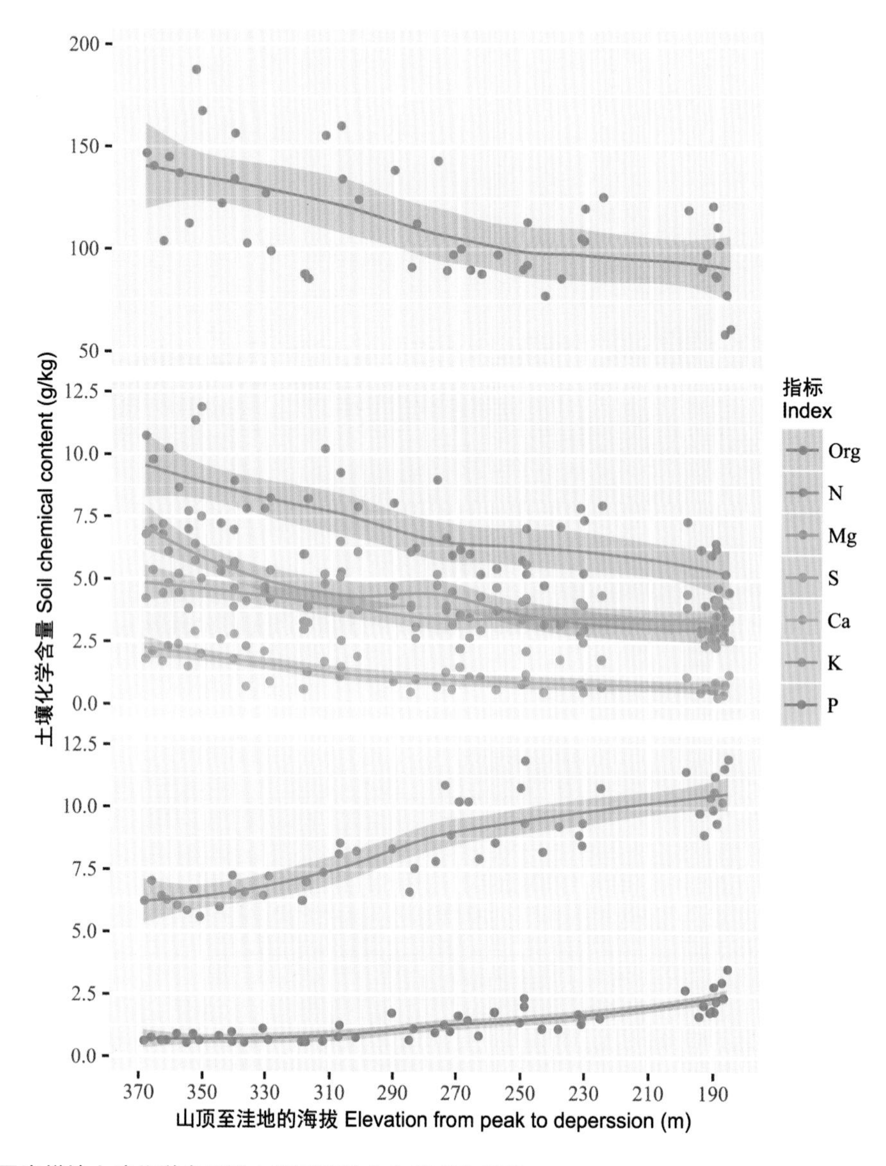

图 6 弄岗样地土壤化学含量沿山顶到洼地方向的变化趋势

Figure 6 Change trends of soil chemical content from peak to depression in the Nonggang plot

Org：有机质 Organic content；N：全氮 Total nitrogen；Mg：镁 Magnesium；S：全硫 Total sulfur；Ca：钙 Calcium；K：全钾 Total potassium；P：全磷 Total phosphorus

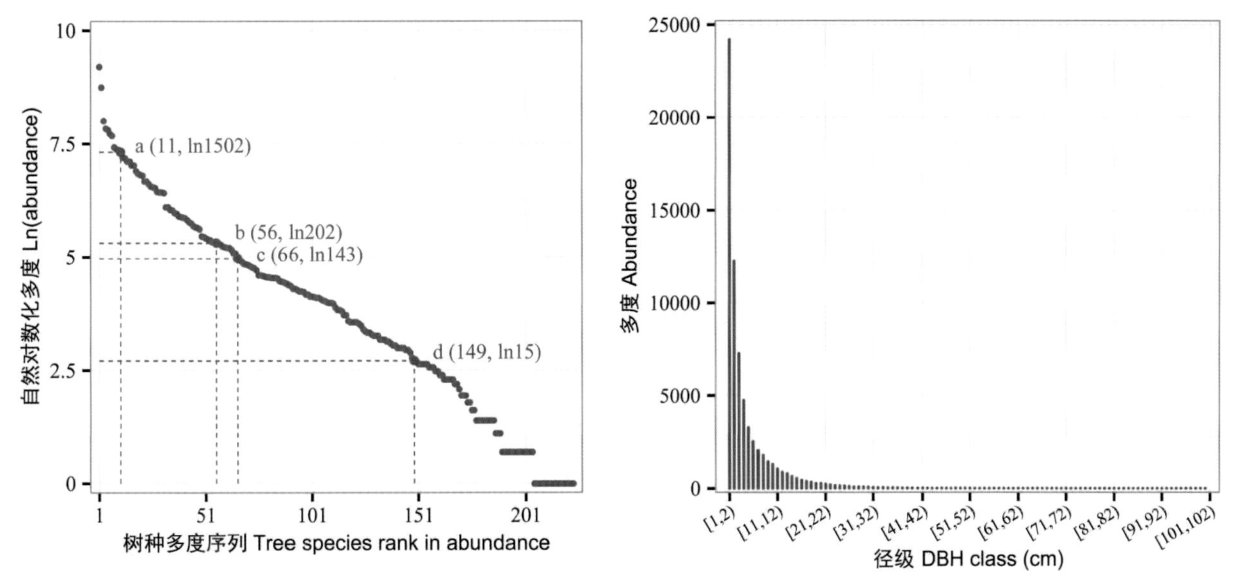

图 7 弄岗样地树种多度序列图和径级分布图

Figure 7 The sequence curve of species abundance and the DBH size distribution of all individuals in the Nonggang plot

2.3 物种组成与群落结构

样地共有 68,010 株胸径（DBH）≥1 cm 的独立木本植物个体（含分枝为 95,471 个体），隶属于 56 科 157 属 223 种[1]。其中，喀斯特特有树种 25,851 株（含分枝为 39,604 个体），隶属于 37 科 57 属 60 种。热带分布的科占总科数的 70.45%，热带分布的属占总属数的 88.39%；主要优势科为大戟科（Euphorbiaceae）、马鞭草科（Verbenaceae）、梧桐科（Sterculiaceae）、桑科（Moraceae）、椴树科（Tiliaceae）、茜草科（Rubiaceae）、楝科（Meliaceae）等（表 1）。山顶优势种为黄梨木（*Boniodendron minus*）、山榄叶柿（*Diospyros siderophylla*）、细叶谷木（*Memecylon scutellatum*）、毛叶铁榄（*Sinosideroxylon pedunculatum* var. *pubifolium*）、清香木（*Pistacia weinmannifolia*）等；山坡优势种为蚬木（*Excentrodendron tonkinense*）、肥牛树（*Cleistanthus petelotii*）、海南椴（*Diplodiscus trichosperma*）、闭花木（*Clcistanthus sumatranus*）、苹婆（*Sterculia monosperma*）、广西牡荆（*Vitex kwangsiensis*）等；洼地优势种广西棋子豆（*Archidendron guangxiensis*）、中国无忧花（*Saraca dives*）、劲直刺桐（*Erythrina stricta*）、日本五月茶（*Antidesma japonicum*）、对叶榕（*Ficus hispida*）等。

样地内少量的常见树种贡献了大部分的个体数量，丰富的偶见树种和稀有树种贡献了少部分的个体数量。其中，有 11 个树种的个体数 >1,500 株，其个体数之和占总个体数的 51.64%；有 56 个树种的个体数 >200 株，占总个体数的 90.19%；有 83 个树种为偶见种，其个体数为 1~10 株/hm^2，占总物种数的 37.22%，占总个体数的 5.2%；有 75 个树种为稀有种，其个体数 ≤1 株/hm^2，占总物种数的 33.63%，占总个体数的 0.35%（图 7 左）。

样地内个体总径级结构呈近似 “L” 形，小径级个体数量最多，之后的个体数量随着径级增加而急剧下降，反映了群落结构较稳定且更新良好等特点。其中，胸径 1~2 cm 的个体有 24,203 株，占总个体数

[1]本书详细描述了样地中 217 个树种，忽略了 6 个未能准确鉴定的树种。

的 35.59%；胸径 2~5 cm 的有 24,262 株，占 35.67%；胸径 5~10 cm 的有 11,040 株，占 16.23%；胸径 10~20 cm 的有 6,593 株，占 9.69%；胸径 20~35 cm 的有 1,670 株，占 2.46%；胸径 35~60 cm 的有 227 株，占 0.33%；胸径大于 60 cm 的有 15 株，占 0.02%（图 7 右）。样地内个体最大胸径为 101 cm，平均胸径为 4.84 cm。

样地内群落结构的空间变异性强烈。山顶周围林高 5~6 m；山坡中部林高 20~30 m，个别树高可达 35 m；洼地周围林高 25~35 m，个别树高可达 45 m。胸径 <20 cm 的个体较广泛地分布于整个样地；胸径 ⩾20 cm 的个体较多分布在山坡中下部，极少分布在山顶周围。以 20 m×20 m 为单位样方的统计结果表明：单位样方的胸高断面积为 0.24~1.80 m^2/400 m^2，平均 0.77 m^2/400 m^2；物种丰富度为 13~56 种/400 m^2，平均 32.07 种/400 m^2；个体多度为 62~420 株/400 m^2，平均 181.36 株/400 m^2。随着“峰丛—洼地”坡面的海拔降低过程，单位样方的胸高断面积呈上升趋势，物种丰富度和个体多度呈下降趋势（图 8）。

表 1 弄岗森林样地重要值排名前 12 位的科

Table 1 Top 12 families with the highest importance values in the Nonggang plot

科名 Family	分布区类型 Areal types	属数 No. of genera	树种数 No. of species	个体数 No. of individual	断面积之和 Basal area (m^2)	重要值 IV
大戟科 Euphorbiaceae	泛热带 Pantropic	21	36	18,203	61.60	19.78
马鞭草科 Verbenaceae	热带亚洲及热带南美间断 Tropical Asia and tropical America disjuncted	4	9	3,649	48.10	8.67
梧桐科 Sterculiaceae	泛热带 Pantropic	4	5	8,047	37.81	8.62
桑科 Moraceae	世界广布 Cosmopolitan	6	18	4,838	16.42	6.36
椴树科 Tiliaceae	泛热带 Pantropic	3	3	2,744	27.39	4.63
茜草科 Rubiaceae	世界广布 Cosmopolitan	9	11	1,800	7.81	3.35
楝科 Meliaceae	世界广布 Cosmopolitan	8	9	2,605	6.82	3.24
番荔枝科 Annonaceae	泛热带 Pantropic	6	7	1,980	2.54	2.55
无患子科 Sapindaceae	泛热带 Pantropic	6	6	1,000	9.86	2.53
蝶形花科 Fabaceae	世界广布 Cosmopolitan	4	4	319	17.22	2.46
柿树科 Ebenaceae	泛热带 Pantropic	1	4	2,520	5.61	2.36
紫葳科 Bignoniaceae	泛热带 Pantropic	4	6	643	8.47	2.03
合计 Total:		76	118	48,348	249.65	66.58
相对百分比 Relative percentage:		48.41%	52.91%	71.09%	75.87%	66.58%

2.3 Species Composition and Community Structure

A total of 68,010 individual trees (95,471 individuals with branch), belonging to 223 species, 157 genera, and 56 families, were recorded in the plot [2]. Among them, 25,851 individual trees (39,604 individuals with branch), belonging to 60 species, 57 genera, and 37 families, are karst endemic tree species. Tropical families account for 70.45% of the total families, and the tropical genera account for 88.39% of the total genera. The most dominant families are Euphorbiaceae, Verbenaceae, Sterculiaceae, Moraceae, Tiliaceae,

[2] 217 species in the plot were described in detail, and 6 species that have not been identified were not included in this book.

Rubiaceae, and Meliaceae. Dominant species of different habitats vary sharply in the plot: *Boniodendron minus*, *Diospyros siderophylla*, *Memecylon scutellatum*, *Sinosideroxylon pedunculatum* var. *pubifolium*, and *Pistacia weinmannifolia* are the dominant species around the mountain peaks; *Excentrodendron tonkinense*, *Cleistanthus petelotii*, *Diplodiscus trichosperma*, *Cleistanthus sumatranus*, *Sterculia monosperma*, and *Vitex kwangsiensis* are the dominant species at the middle slopes; *Archidendron guangxiensis*, *Saraca dives*, *Erythrina stricta*, *Antidesma japonicum*, and *Ficus hispida* are the dominant species at the valley bottoms.

Eleven species with more than 1,500 individuals account for 51.64% of the total individuals; 58 species with more than 200 individuals account for 90.19%; 83 species with 1–10 individuals per hectare, which are considered as occasional species, account for 37.22% of the total species and 5.2% of the total individuals; 75 species with less than one individual per hectare, which are considered as rare species, account for 33.63% of the total species and 0.35% of the total individuals(Figure 7 left) .

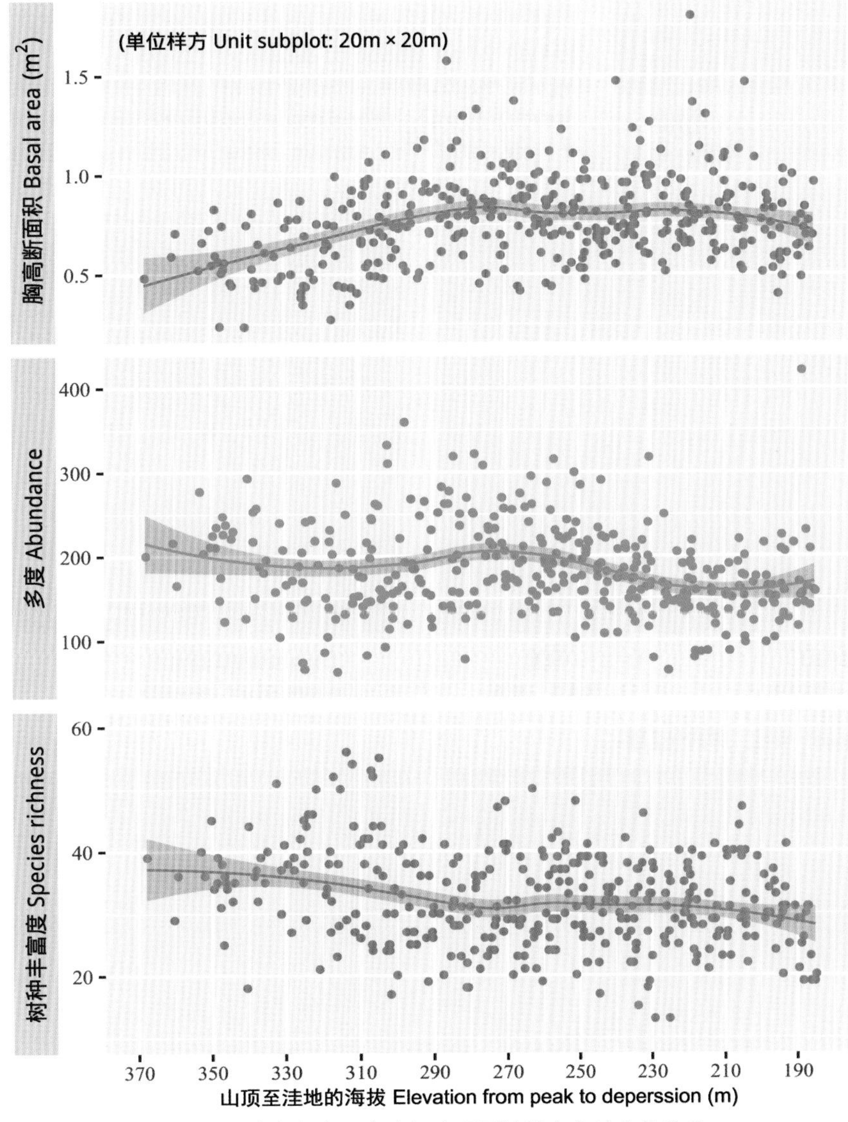

图 8 弄岗样地胸高断面积、树种丰富度和多度沿山顶到洼地方向的变化趋势

Figure 8 Change trends of basal area, species richness and abundance from peak to depression in the Nonggang plot

The DBH size distribution of the total individuals exhibits an approximate L-shaped pattern in the plot. Most of the individuals found in the plot belong to the small-sized diameter class, then the numbers of individuals in different diameter classes decrease rapidly as the diameter class size increases, indicating a stable structure and good regeneration of the forest communities in the plot. 24,203 individuals with DBH between 1–2 cm account for 35.59% of total individuals in the plot, 24,262 individuals with DBH between 2–5 cm account for 35.67%, 11,040 individuals with DBH between 5–10 cm account for 16.23%, 6,593 individuals with DBH between 10–20 cm account for 9.69%, 1,670 individuals with DBH between 20–35 cm account for 2.46%, 227 individuals with DBH between 35–60 cm account for 0.33%, and 15 individuals with DBH more than 60 cm account for 0.02%. The largest DBH of individuals is 101 cm, and mean value of DBH is 4.84 cm in the plot (Figure 7 right) .

Community structures also show strong spatial heterogeneity in the plot. The forest canopy heights are 5–6 m around the peaks; the forest canopy heights are 20–30 m, and a few tree heights reach 35 m at the middle slopes; the forest canopy heights are 25–35 m, and a few tree heights reach 45 m at the valley bottoms. The tree individuals with DBH<20 cm distributed evenly over the entire plot, while the tree individuals with DBH⩾20 cm show higher density at the valley bottoms. The statistical results based on subplots of 20 m×20 m show that: the basal area range in 0.24–1.80 m^2/400 m^2, averaging 0.77 m^2/400 m^2; species richness range in 13–56 species/400 m^2, averaging 32.07 species/400 m^2; individual abundance range in 62–420 individuals/400 m^2, averaging 181.36 individuals/400 m^2. As the altitude of "peak-depression" slope location decreases, the basal area per unit subplot shows an increasing trend, the species richness and individual abundance show decreasing trends (Figure 8).

3 弄岗样地的树种及其分布格局 Tree Species and Their Distribution Patterns in Nonggang plot

1 显脉木兰

xiǎn mài mù lán | Long-leaf Magnolia

Lirianthe fistulosa (Finet et Gagnep.) N. H. Xia et C. Y. Wu
木兰科 Magnoliaceae

代码（SpCode）= LIRFIS
个体数（Individual number/15 hm^2）= 97
最大胸径（Max DBH）= 7.8 cm
重要值排序（Importance value rank）= 94

常绿灌木或小乔木，高可达 11 m。小枝多少被短柔毛，老时柔毛脱落，托叶痕几达叶柄的顶端。叶无毛，叶脉在背部明显突出。花梗开花时下弯，结果时近直立。果长圆形，长约 5 cm。花期 4~7 月，果期 7~10 月。

Evergreen shrubs or small trees, up to 11 m tall. Branchlets pubescent more or less, order branches glabrous, stipule vestige almost to the petiole apex. Leaves glabrous, dorsal nerves obvious salient. Pedicel declined when efflorescence, almost erect when fruit. Fruit oblong, about 5 cm long. Fl. Apr.–Jul., fr. Jul.–Oct..

果枝 Fruiting branches
摄影：黄俞淞 Photo by: Huang Yusong

叶和花蕾 Leaves and flower bud
摄影：丁涛 Photo by: Ding Tao

花 Flower
摄影：黄俞淞 Photo by: Huang Yusong

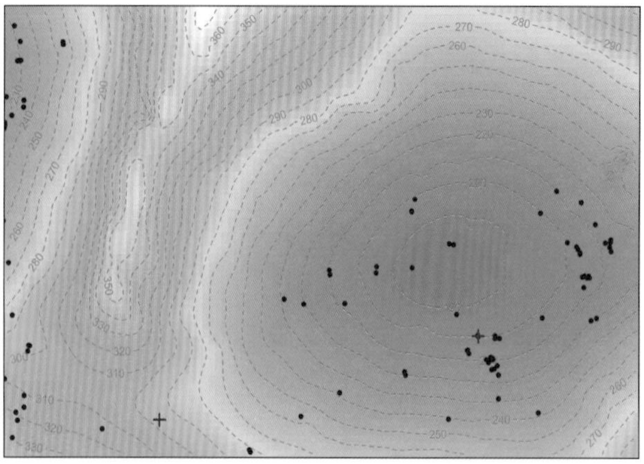

• 1~5 cm DBH + 5~20 cm DBH ○ ≥20 cm DBH
个体分布图 Distribution of individuals

径级分布表 DBH class

胸径区间 (Diameter class) (cm)	个体数 (No. of individuals in the plot)	比例 (Proportion) (%)
1~2	68	70.10
2~5	27	27.84
5~10	2	2.06
10~20	0	0.00
20~35	0	0.00
35~60	0	0.00
≥60	0	0.00

2 石密

shí mì | Soft Alphonsea

Alphonsea mollis Dunn
番荔枝科 Annonaceae

代码（SpCode）= ALPMOL
个体数（Individual number/15 hm^2）= 65
最大胸径（Max DBH）= 17.9 cm
重要值排序（Importance value rank）= 101

常绿乔木，高达 20 m。树皮暗灰褐色。叶纸质，叶背明显被柔毛。叶柄短，被柔毛。花淡黄色，单生或双生。果成熟时黄色，单个或双个，被黄褐色柔毛。花期早春，果期 6~8 月。

Evergreen trees, up to 20 m tall, Bark fuscous. Leaves chartaceous, leaf dorsal obvious pilose. Petiole short, pilose. Flowers xanthic, solitary or twins. Fruit luteous at maturitye, solitary or twins, pilose fulvous. Fl. prevernal, fr. Jun.–Aug..

树干 Trunk
摄影：黄俞淞 Photo by：Huang Yusong

枝叶 Branches and leaves
摄影：黄俞淞 Photo by：Huang Yusong

果 Fruits
摄影：黄俞淞 Photo by：Huang Yusong

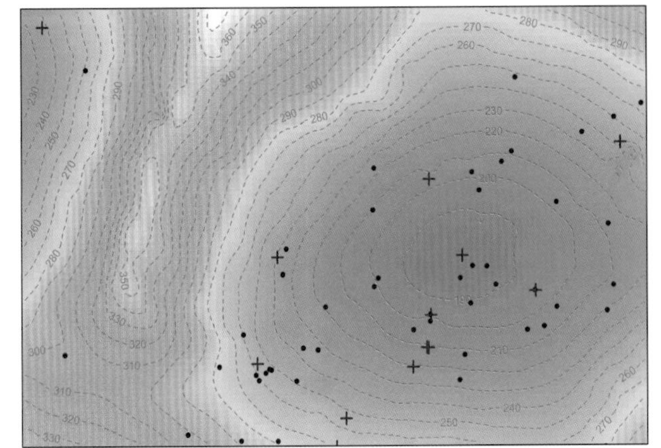

• 1~5 cm DBH + 5~20 cm DBH ○ ≥20 cm DBH

个体分布图 Distribution of individuals

径级分布表 DBH class

胸径区间 (Diameter class) (cm)	个体数 (No. of individuals in the plot)	比例 (Proportion) (%)
1~2	26	40.00
2~5	26	40.00
5~10	9	13.85
10~20	4	6.15
20~35	0	0.00
35~60	0	0.00
≥60	0	0.00

3 藤春

téng chūn | One-pistil Alphonsea

Alphonsea monogyna Merr. et Chun
番荔枝科 Annonaceae

代码（SpCode）= ALPMON
个体数（Individual number/15 hm^2）= 387
最大胸径（Max DBH）= 25.1 cm
重要值排序（Importance value rank）= 37

常绿乔木，高可达 12 m。小枝被疏柔毛，叶纸质或近革质，顶端急尖或渐尖，两面无毛，干后灰白色。花黄色，花梗被锈色短柔毛。果近圆球形或椭圆形，具不明显的小瘤体。花期 1~9 月，果期 9 月至翌年春季。

Evergreen trees, to 12 m tall. Branchlets pubescent, leaves chartaceous or almost coreaceous, apex acute or acuminate, both surfaces glabrous, canescent when dry. Flowers yellow, pedicel rubiginose pubescent. Fruit suborbicular or elliptic, not obvious nodulose. Fl. Jan.–Sep., fr. Sep.–springtime of next year.

树干 Trunk
摄影：黄俞淞 Photo by：Huang Yusong

果枝 Fruiting branch
摄影：刘晟源 Photo by：Liu Shengyuan

花 Flower
摄影：黄俞淞 Photo by：Huang Yusong

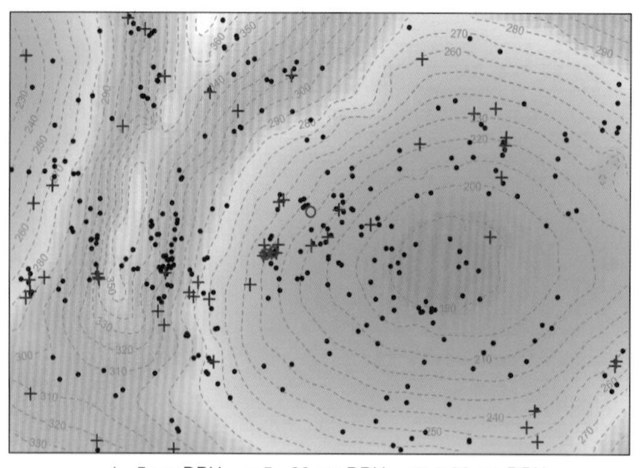

• 1~5 cm DBH + 5~20 cm DBH ○ ≥20 cm DBH
个体分布图 Distribution of individuals

径级分布表 DBH class

胸径区间 (Diameter class) (cm)	个体数 (No. of individuals in the plot)	比例 (Proportion) (%)
1~2	166	42.89
2~5	164	42.38
5~10	42	10.85
10~20	12	3.10
20~35	3	0.78
35~60	0	0.00
≥60	0	0.00

4 假鹰爪

jiǎ yīng zhuǎ | Chinese Desmos

Desmos chinensis Lour.
番荔枝科 Annonaceae

代码（SpCode）= DESCHI
个体数（Individual number/15 hm^2）= 207
最大胸径（Max DBH）= 9.3 cm
重要值排序（Importance value rank）= 62

常绿直立或攀缘灌木。树皮粗糙，具纵条纹，有灰白色凸起的皮孔。叶基部圆形或稍偏斜，叶背常白粉状。花淡黄色，单朵与叶对生或互生。果有柄，念珠状。花期 4~10 月，果期 6 月至翌年春季。

Evergreen erect or climbing shrubs. Bark asperous, longitudinal striate, pallid salient lenticel. Leaf base rounded or little oblique, leaf dorsal glaucous. Flower ochroleucous, opposite or alternate with leaves. Fruit cauliferous, moniliform. Fl. Apr.–Oct., fr. Jun.–springtime of next year.

果枝 Fruiting branches
摄影：黄俞淞 Photo by：Huang Yusong

果序 Infructescence
摄影：黄俞淞 Photo by：Huang Yusong

花 Flower
摄影：黄俞淞 Photo by：Huang Yusong

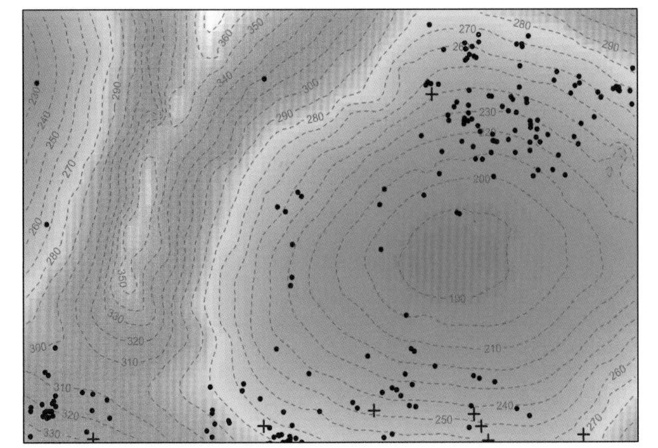

● 1~5 cm DBH ＋ 5~20 cm DBH ○ ≥20 cm DBH
个体分布图 Distribution of individuals

径级分布表 DBH class

胸径区间 (Diameter class) (cm)	个体数 (No. of individuals in the plot)	比例 (Proportion) (%)
1~2	144	69.57
2~5	55	26.57
5~10	8	3.86
10~20	0	0.00
20~35	0	0.00
35~60	0	0.00
≥60	0	0.00

5 田方骨 tián fāng gǔ | Vietnam Goniothalamus

Goniothalamus donnaiensis Finet et Gagnep.
番荔枝科 Annonaceae

代码（SpCode）= GONDON
个体数（Individual number/15 hm^2）= 74
最大胸径（Max DBH）= 14.5 cm
重要值排序（Importance value rank）= 117

常绿灌木或小乔木，高可达 4 m。枝条灰黑色，有不规则纵裂纹，幼枝、叶背、叶柄和果均密被红褐色短硬毛。叶基部楔形，顶端具尾尖。花单朵腋生。果 4~12 个聚生在一起，卵状长圆形。花期 5~9 月，果期 8~12 月。

Evergreen shurbs or small trees, to 4 m tall. Branches cinereous, with unequal longitudinal stripe, branchlet, leaf dorsal, petiole and fruit coarctate fulvous barbellate. Base cuneate, apex tailed cusp. Flowers solitary axillary, fruit 4–12 aggregate, ovate oblong. Fl. May–Sep., fr. Aug.–Dec..

果枝 Fruiting branch
摄影：黄俞淞 Photo by：Huang Yusong

果序 Infructescence
摄影：黄俞淞 Photo by：Huang Yusong

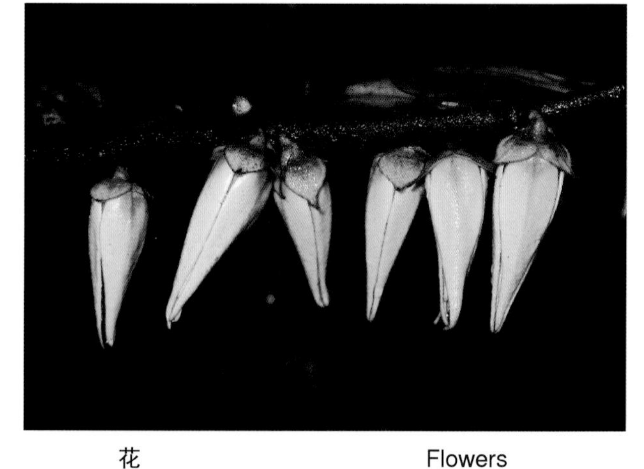

花 Flowers
摄影：黄俞淞 Photo by：Huang Yusong

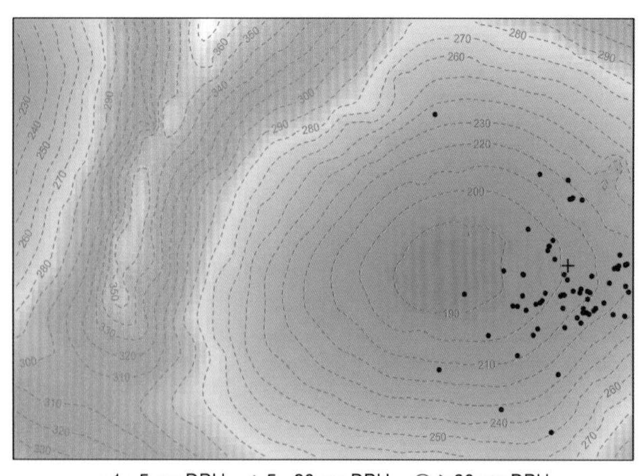

• 1~5 cm DBH　+ 5~20 cm DBH　○ ≥20 cm DBH
个体分布图 Distribution of individuals

径级分布表 DBH class

胸径区间 (Diameter class) (cm)	个体数 (No. of individuals in the plot)	比例 (Proportion) (%)
1~2	66	89.19
2~5	7	9.46
5~10	0	0.00
10~20	1	1.35
20~35	0	0.00
35~60	0	0.00
≥60	0	0.00

6 野独活

yě dú huó | Chun's Miliusa

Miliusa chunii W. T. Wang
番荔枝科 Annonaceae

代码（SpCode）＝ MILCHU
个体数（Individual number/15 hm^2）＝ 14
最大胸径（Max DBH）＝ 18.3 cm
重要值排序（Importance value rank）＝ 156

常绿灌木，高 2~5 m。小枝稍被伏贴短柔毛。叶膜质，基部宽楔形或圆形，偏斜，侧脉每边 10~12 条。花单生叶腋内，花梗丝状，长 4~6.5 cm。果圆球状，内有种子 1~3 颗。花期 4~7 月，果期 7 月至翌年春季。

Evergreen trees, to 2–5 m tall. Branchlets slightly pubescent. Leaf membranous, base cuneate or rounded, oblique, lateral veins 10–12 on each side of midvein. Inflorescences axillary, 1-flowered, Pedicel filiform, 4–6.5 cm long. Monocarp globose, Seeds 1–3 per monocarp. Fl. Apr.–Jul., fr. Jul.–springtime of next year.

花枝 Flowering branches
摄影：黄俞淞 Photo by：Huang Yusong

果序 Infructescence
摄影：丁涛 Photo by：Ding Tao

花 Flowers
摄影：黄俞淞 Photo by：Huang Yusong

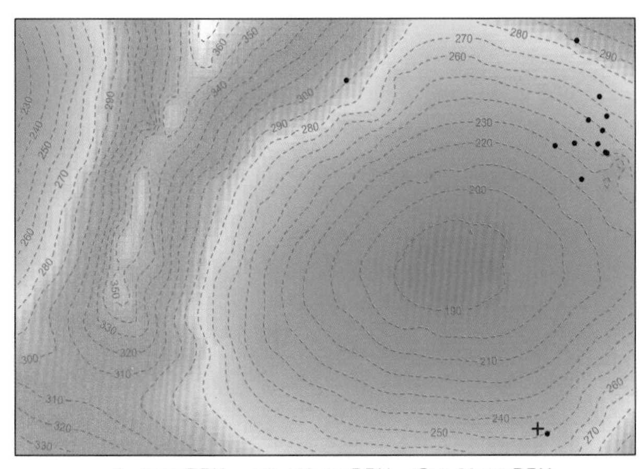

• 1~5 cm DBH + 5~20 cm DBH ○ ≥20 cm DBH
个体分布图 Distribution of individuals

径级分布表 DBH class

胸径区间 (Diameter class) (cm)	个体数 (No. of individuals in the plot)	比例 (Proportion) (%)
1~2	8	57.14
2~5	5	35.71
5~10	0	0.00
10~20	1	7.14
20~35	0	0.00
35~60	0	0.00
≥60	0	0.00

7 山蕉

shān jiāo | Maingay's Mitrephora

Mitrephora maingayi Hook. f. et Thomson
番荔枝科 Annonaceae

代码（SpCode）= MITMAI
个体数（Individual number/15 hm^2）= 10
最大胸径（Max DBH）= 7.4 cm
重要值排序（Importance value rank）= 174

常绿乔木，高 6~20 m。树皮灰黑色。叶革质，被疏柔毛，老渐无毛。花两性，初时白色，后变黄色，有红色斑点，单生或数朵生于被锈色柔毛的总花梗上，总花梗与叶对生，或腋外生。果卵状或圆柱状，被锈色短柔毛。花期 2~8 月，果期 6~12 月。

Evergreen trees, to 6–20 m tall. Bark cinereous. Leaves coreaceous, puberulent, grabrous when older. Flowers bisexual, from white to yellow, with red macula, solitary or multiflorous on rubiginose pilose pedicel, oppositiflorous, or extra-axillary. Fruit ovate or terete, rubiginose pubescent. Fl. Feb.–Aug., fr. Jun.–Dec..

花枝 Flowering branch
摄影：黄俞淞 Photo by: Huang Yusong

花 Flower
摄影：黄俞淞 Photo by: Huang Yusong

叶背 Leaf back
摄影：丁涛 Photo by: Ding Tao

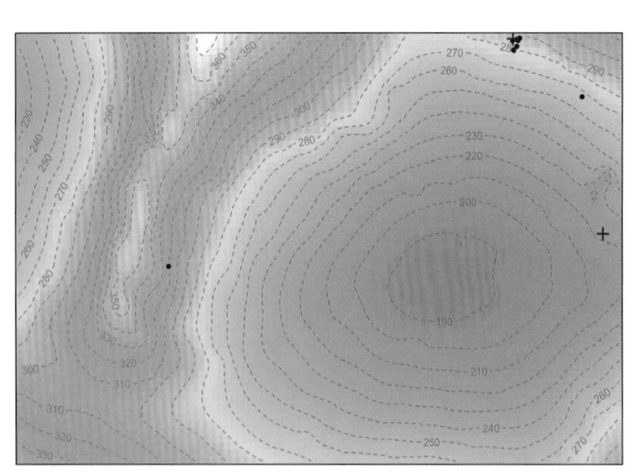

• 1~5 cm DBH + 5~20 cm DBH ○ ≥20 cm DBH
个体分布图 Distribution of individuals

径级分布表 DBH class

胸径区间 (Diameter class) (cm)	个体数 (No. of individuals in the plot)	比例 (Proportion) (%)
1~2	6	60.00
2~5	2	20.00
5~10	2	20.00
10~20	0	0.00
20~35	0	0.00
35~60	0	0.00
≥60	0	0.00

8 广西澄广花

guǎng xī chéng guǎng huā | Guangxi Orophea

Orophea anceps Pierre
番荔枝科 Annonaceae

代码（SpCode）= OROANC
个体数（Individual number/15 hm^2）= 1223
最大胸径（Max DBH）= 22.8 cm
重要值排序（Importance value rank）= 18

常绿灌木或小乔木，高达 8 m。枝条灰褐色，嫩枝被柔毛。叶基部两侧不对称，偏斜。花淡红色，单朵腋生，内轮花瓣基部具很窄的爪。果圆球形，无毛。花期 7~9 月，果期 8~11 月。

Evergreen shrubs or small trees, to 8 m tall. Branches cinereous, branchlet pilose. Base aequilateral, oblique. Flowers mauve, solitary axillary, including petal base with narrow claw. Fruit round, glabrous. Fl. Jul.–Sep., fr. Aug.–Nov..

植株 Whole plant
摄影：黄俞淞 Photo by：Huang Yusong

花 Flower
摄影：丁涛 Photo by：Ding Tao

果枝 Fruiting branch
摄影：黄俞淞 Photo by：Huang Yusong

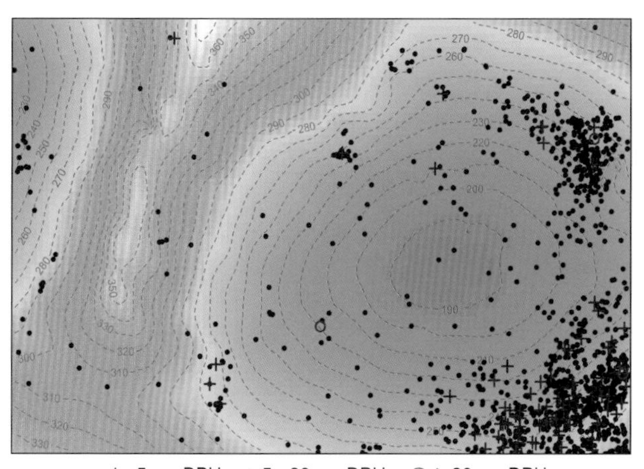

● 1~5 cm DBH　+ 5~20 cm DBH　○ ≥20 cm DBH
个体分布图 Distribution of individuals

径级分布表 DBH class

胸径区间 (Diameter class) (cm)	个体数 (No. of individuals in the plot)	比例 (Proportion) (%)
1~2	527	43.09
2~5	604	49.39
5~10	88	7.20
10~20	1	0.08
20~35	3	0.25
35~60	0	0.00
≥60	0	0.00

9 岩樟 yán zhāng | Rockdwelling Cinnamon

Cinnamomum saxatile H. W. Li
樟科 Lauraceae

代码（SpCode）= CINSAX
个体数（Individual number/15 hm^2）= 69
最大胸径（Max DBH）= 33.7 cm
重要值排序（Importance value rank）= 88

常绿乔木，高可达 15 m。树枝干后变棕黑色，无毛，嫩枝通常压扁，被棕色微茸毛。叶椭圆形或长椭圆状卵形，叶脉 5~7 对，和小叶脉交织成网状，呈蜂窝状。圆锥花序近顶生，3~6 cm，具花 6~15 朵，花绿色；果卵圆形，成熟后棕黑色。花期 4~5 月，果期 10 月。

Evergreen trees, to 15 m tall. Branchlets black-brown when dry, glabrous, young branchlets consipicuously compressed, brownish puberulent. Leave blade oblong or ovate-oblong, veins 5–7 pairs, conspicuously foveolate.Panicle subterminal, 3–6 cm, 6–15 flowered, flowers green. Fruit ovate, black-brown when maturity. Fl. Apr.–May, fr. Oct..

枝叶 Branch and leaves
摄影：黄俞淞 Photo by：Huang Yusong

果枝 Fruiting branches
摄影：黄俞淞 Photo by：Huang Yusong

果 Fruit
摄影：黄俞淞 Photo by：Huang Yusong

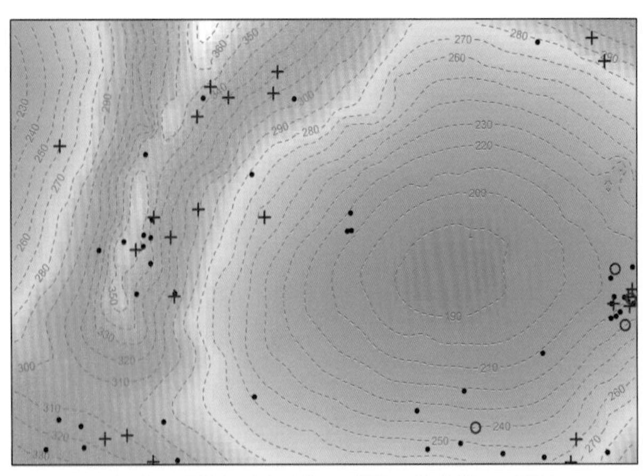

• 1~5 cm DBH + 5~20 cm DBH ○ ≥20 cm DBH
个体分布图 Distribution of individuals

径级分布表 DBH class

胸径区间 (Diameter class) (cm)	个体数 (No. of individuals in the plot)	比例 (Proportion) (%)
1~2	20	28.99
2~5	22	31.88
5~10	12	17.39
10~20	11	15.94
20~35	4	5.80
35~60	0	0.00
≥60	0	0.00

10 南烛厚壳桂 nán zhú hòu ké guì | Pitted-leaf Cryptocarya

Cryptocarya lyoniifolia S. Lee et F. N. Wei
樟科 Lauraceae

代码（SpCode）= CRYLYO
个体数（Individual number/15 hm^2）= 286
最大胸径（Max DBH）= 48.2 cm
重要值排序（Importance value rank）= 46

常绿乔木，高达 20 m。小枝有纵条纹，无毛。叶基部楔形或近圆形，侧脉稀疏，每边 3~5 条，在叶背构成蜂窝状，两面无毛。果序顶生或腋生，被短柔毛；果球形，无毛，表面具皱纹。果期 10~12 月，少数延长到翌年 1 月。

Evergreen trees, to 20 m tall. Branchlets longitudinal, grabrous. Leaf base cuneate or suborbicular, each lateral veins sparse 3–5, dorsal leaf favose, both faces grabrous. Fruit globose, glabrous, rugose. Fr. Oct.–Dec. (or few to Jan. of next year).

幼苗 Seedling
摄影：黄俞淞 Photo by：Huang Yusong

叶 Leaves
摄影：黄俞淞 Photo by：Huang Yusong

果枝 Fruiting branches
摄影：黄俞淞 Photo by：Huang Yusong

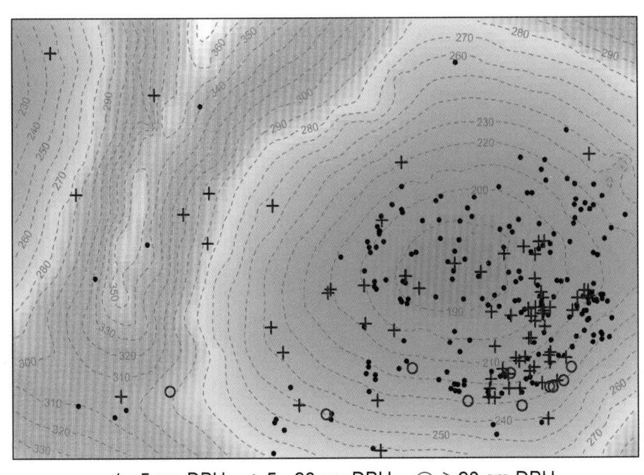

• 1~5 cm DBH + 5~20 cm DBH ○ ≥20 cm DBH
个体分布图 Distribution of individuals

径级分布表 DBH class

胸径区间 (Diameter class) (cm)	个体数 (No. of individuals in the plot)	比例 (Proportion) (%)
1~2	103	36.01
2~5	100	34.97
5~10	54	18.88
10~20	19	6.64
20~35	5	1.75
35~60	5	1.75
≥60	0	0.00

11 蜂窝木姜子

fēng wō mù jiāng zǐ | Pitted-leaf Litse

Litsea foveola Kosterm.
樟科 Lauraceae

代码（SpCode）= LITFOV
个体数（Individual number/15 hm^2）= 362
最大胸径（Max DBH）= 12.5 cm
重要值排序（Importance value rank）= 41

常绿灌木或小乔木，高 3~5 m。幼枝灰褐色，被灰黄色长柔毛。叶具明显的蜂窝状小穴。伞形花序常 2 个簇生于枝顶叶腋，每一花序有花 4~5 朵。果球形。花期 7~8 月，果期 12 月。

Evergreen shrubs or trees, 3–5 m tall. Branchlet cinereous, densely gray-yellow villous. Leaf conspicuously foveolate adaxially. Umbels axillary, often 2-clustered at apex of branchlet, 4–5-flowered. Fruit globose. Fl. Jul.–Aug., fr. Dec..

花 Flower
摄影：黄俞淞 Photo by：Huang Yusong

果 Fruit
摄影：黄俞淞 Photo by：Huang Yusong

果枝 Fruiting branch
摄影：黄俞淞 Photo by：Huang Yusong

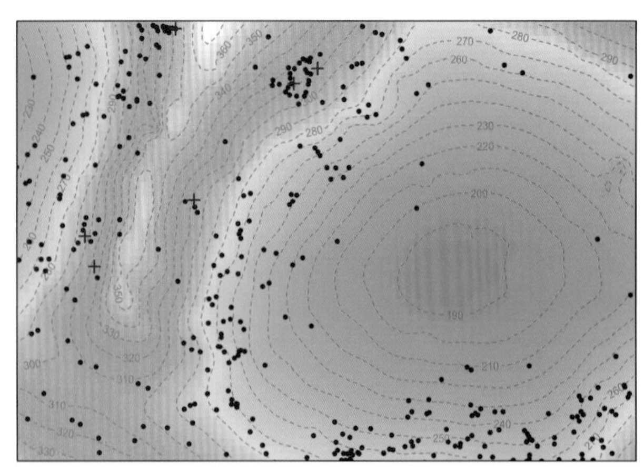

• 1~5 cm DBH + 5~20 cm DBH ○ ≥20 cm DBH
个体分布图 Distribution of individuals

径级分布表 DBH class

胸径区间 (Diameter class) (cm)	个体数 (No. of individuals in the plot)	比例 (Proportion) (%)
1~2	261	72.10
2~5	95	26.24
5~10	5	1.38
10~20	1	0.28
20~35	0	0.00
35~60	0	0.00
≥60	0	0.00

12 潺槁木姜子 chán gāo mù jiāng zǐ | Soft Bollygum

Litsea glutinosa (Lour.) C. B. Rob.
樟科 Lauraceae

代码（SpCode）= LITGLU
个体数（Individual number/15 hm^2）= 10
最大胸径（Max DBH）= 14.4 cm
重要值排序（Importance value rank）= 159

常绿或落叶乔木，高 3~15 m。树皮灰色或灰褐色，幼枝被灰黄色绒毛。叶互生，幼时两面被毛，叶柄长有灰黄色茸毛。伞形花序生于小枝上部叶腋。果球形。花期 5~6 月，果期 9~10 月。

Evergreen or deciduous trees, 3–15 m tall. Bark gray or taupe, young branchlets gray-yellow tomentose. Leaves alternate, tomentose on both when young, petiole gray-yellow tomentose. Umbels axillary, often at apex of branchlet. Fruit globose. Fl. May–Jun., fr. Sep.–Oct..

花枝 Flowering branch
摄影：黄俞淞 Photo by：Huang Yusong

果 Fruits
摄影：黄俞淞 Photo by：Huang Yusong

果枝 Fruiting branch
摄影：黄俞淞 Photo by：Huang Yusong

• 1~5 cm DBH + 5~20 cm DBH ○ ≥20 cm DBH
个体分布图 Distribution of individuals

径级分布表 DBH class

胸径区间 (Diameter class) (cm)	个体数 (No. of individuals in the plot)	比例 (Proportion) (%)
1~2	1	10.00
2~5	5	50.00
5~10	1	10.00
10~20	3	30.00
20~35	0	0.00
35~60	0	0.00
≥60	0	0.00

13 毛黄椿木姜子 máo huáng chūn mù jiāng zǐ | Oblong Varied Litse

Litsea variabilis var. *oblonga* Lec.
樟科 Lauraceae

代码（SpCode）= LITVAR
个体数（Individual number/15 hm^2）= 24
最大胸径（Max DBH）= 9.8 cm
重要值排序（Importance value rank）= 133

常绿灌木或乔木，高达 15 m。树皮灰色，灰褐色或黑褐色。叶对生或近对生，兼有互生，叶下面和叶柄均密被灰黄色贴伏柔毛。伞形花序常 3~8 个集生叶腋，具 3 朵花。果球形，成熟时黑色。花期 5~11 月，果期 9 月至翌年 5 月。

Evergreen shrubs or trees, up to 15 m tall. Bark gray, taupe or black brown. Leaves opposite or sub opposite, rarely alternate, both leaves abaxial and petiole densely gray-yellow appressed pubescent. Umbels often in cluster of 3–8, axillary, 3-flowered. Fruit globose, black at maturity. Fl. May–Nov., fr. Sep.–May of next year.

小枝 Branchlet
摄影：黄俞淞 Photo by：Huang Yusong

果序 Infructescence
摄影：黄俞淞 Photo by：Huang Yusong

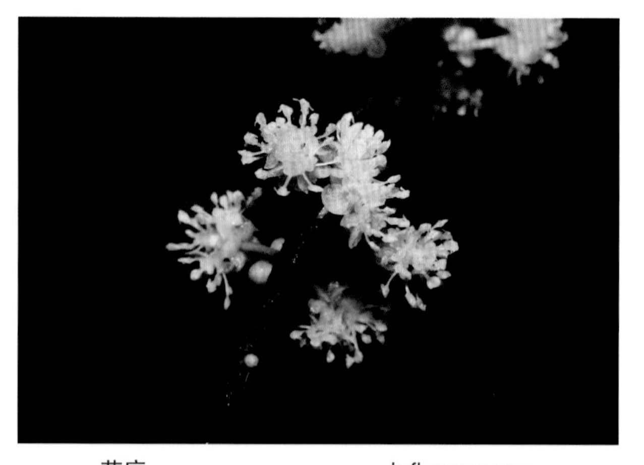

花序 Inflorescence
摄影：黄俞淞 Photo by：Huang Yusong

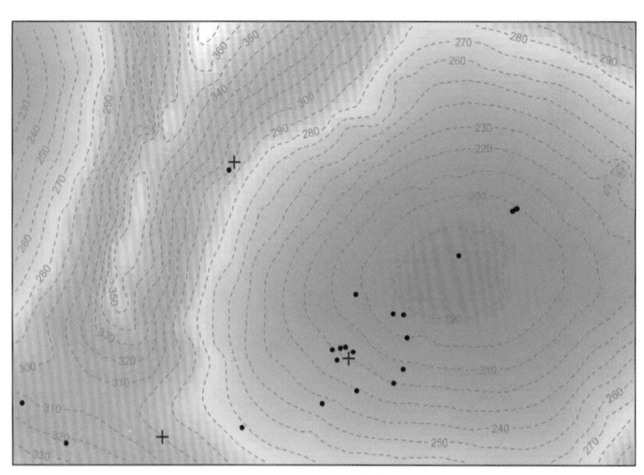

• 1~5 cm DBH + 5~20 cm DBH ○ ≥20 cm DBH
个体分布图 Distribution of individuals

径级分布表 DBH class

胸径区间 (Diameter class) (cm)	个体数 (No. of individuals in the plot)	比例 (Proportion) (%)
1~2	14	58.33
2~5	7	29.17
5~10	3	12.50
10~20	0	0.00
20~35	0	0.00
35~60	0	0.00
≥60	0	0.00

14 海南新木姜子 hǎi nán xīn mù jiāng zǐ | Hainan Newlitse

Neolitsea hainanensis Yen C. Yang et P. H. Huang
樟科 Lauraceae

代码（SpCode）= NEOHAI
个体数（Individual number/15 hm^2）= 5
最大胸径（Max DBH）= 10.6 cm
重要值排序（Importance value rank）= 176

乔木或小乔木，高达 10 m。树皮灰褐色。叶近轮生或互生，椭圆形或圆状椭圆形，两面无毛，均有明显的蜂窝状小穴，离基三出脉。伞形花序 1 至多个簇生叶腋或枝侧。果球形，果托近于扁平盘状，常宿存有花被片。花期 11 月，果期翌年 7~8 月。

Trees or small trees, up to 10 m tall. Bark gray brown. Leaves subverticillate or alternate, elliptic or rounded-elliptic, glabrous and distinctly foveolate on both surfaces, triplinerved. Umbels solitary or fascicled, axillary or lateral. Fruit globose, seated on nearly flat discoid perianth tube, perianth segments often persistent. Fl. Nov., fr. Jul.–Aug. of next year.

枝叶 Branch and leaves
摄影：黄俞淞 Photo by：Huang Yusong

花蕾 Flower buds
摄影：黄俞淞 Photo by：Huang Yusong

叶 Leaves
摄影：黄俞淞 Photo by：Huang Yusong

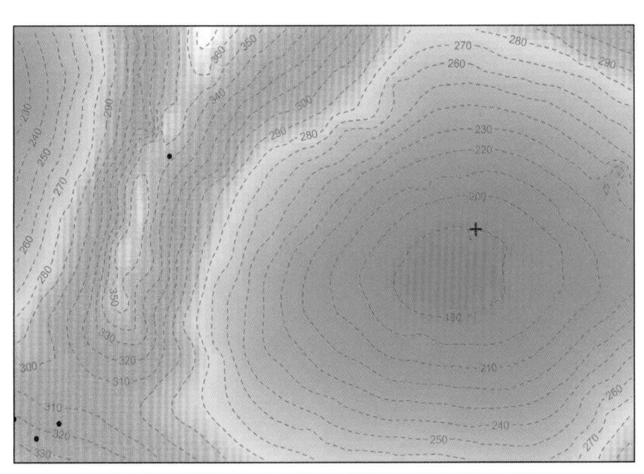

• 1~5 cm DBH + 5~20 cm DBH ○ ≥20 cm DBH
个体分布图 Distribution of individuals

径级分布表 DBH class

胸径区间 (Diameter class) (cm)	个体数 (No. of individuals in the plot)	比例 (Proportion) (%)
1~2	3	60.00
2~5	1	20.00
5~10	0	0.00
10~20	1	20.00
20~35	0	0.00
35~60	0	0.00
≥60	0	0.00

15 石山楠 shí shān nán | Limy Nan

Phoebe calcarea S. Lee et F. N. Wei
樟科 Lauraceae

代码（SpCode）= PHOCAL
个体数（Individual number/15 hm^2）= 21
最大胸径（Max DBH）= 24.3 cm
重要值排序（Importance value rank）= 130

常绿乔木，高 4~15 m。树皮灰白色。叶披针形或椭圆状披针形，先端渐尖或尾状渐尖，尖头常作镰状，嫩叶两面常带紫红色。圆锥花序数个，顶生，花梗通常被白粉。果卵形，先端常有短喙。花期 4~5 月，果期 7~9 月。

Evergreen trees, 4–15 m tall. Bark gray-white. Leaves lanceolate or elliptic lanceolate, apex acuminate or long acuminate, often falciform, both faces often aubergine when young. Panicles many, terminal, pedicel often with whiting. Fruit ovoid, apex often whit short beak. Fl. Apr.–May, fr. Jul.–Sep..

果枝 Fruiting branches
摄影：黄俞淞 Photo by：Huang Yusong

果 Fruits
摄影：黄俞淞 Photo by：Huang Yusong

叶 Leaves
摄影：黄俞淞 Photo by：Huang Yusong

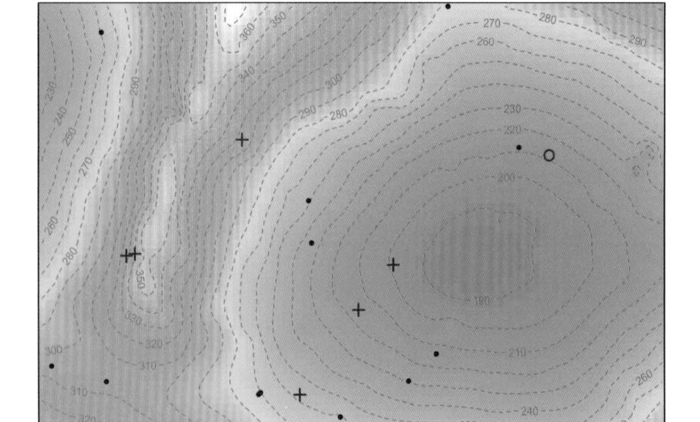
• 1~5 cm DBH + 5~20 cm DBH ○ ≥20 cm DBH
个体分布图 Distribution of individuals

径级分布表 DBH class

胸径区间 (Diameter class) (cm)	个体数 (No. of individuals in the plot)	比例 (Proportion) (%)
1~2	7	33.33
2~5	7	33.33
5~10	4	19.05
10~20	2	9.52
20~35	1	4.76
35~60	0	0.00
≥60	0	0.00

16 无柄山柑

wǔ bǐng shān gān | Subsessile Caper

Capparis subsessilis B. S. Sun
白花菜科 Cleomaceae

代码（SpCode）= CAPSUB
个体数（Individual number/15 hm^2）= 3
最大胸径（Max DBH）= 4.2 cm
重要值排序（Importance value rank）= 189

常绿灌木，高约 3 m。细枝干后浅绿色，光滑无毛，具向上弯曲的小皮刺。叶柄 1~2 mm，叶片椭圆形至倒卵状椭圆形，两面无毛，叶面干后草黄色，叶基部心形，顶端具 0.7~1.6 cm 尾尖。花 1~2 朵腋生。果近球形，略偏斜。果期 8~10 月。

Evergreen shrubs, ca. 3 m tall. Twigs pale green when dry, glabrous, with small ascending stipular spines. Petiole 1–2 mm, leaf blade elliptic to slightly obovate-elliptic, both surfaces glabrous, adaxially grass-yellow when dry, base cordate, apex with a 0.7–1.6 cm tip. Inflorescences superaxillary of 1–2 flowers. Fruit subglobose but slightly oblate. Fr. Aug.–Oct..

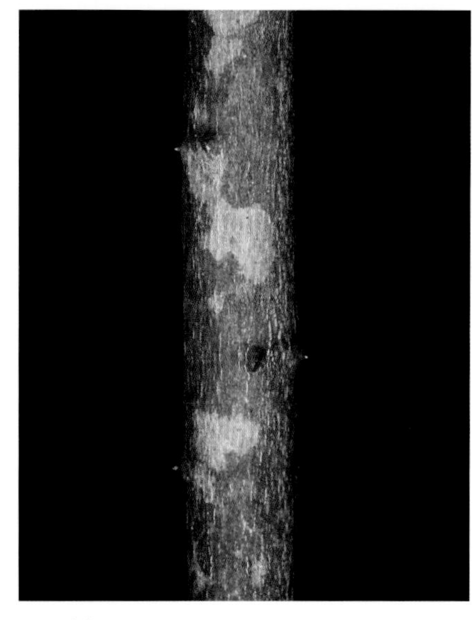

树干 Trunk
摄影：黄俞淞 Photo by: Huang Yusong

枝叶 Branch and leaves
摄影：黄俞淞 Photo by: Huang Yusong

果 Fruits
摄影：黄俞淞 Photo by: Huang Yusong

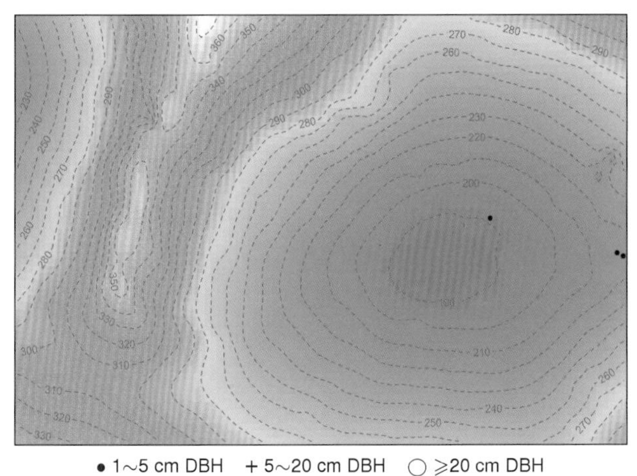

• 1~5 cm DBH + 5~20 cm DBH ○ ≥20 cm DBH
个体分布图 Distribution of individuals

径级分布表 DBH class

胸径区间 (Diameter class) (cm)	个体数 (No. of individuals in the plot)	比例 (Proportion) (%)
1~2	0	0.00
2~5	3	100.00
5~10	0	0.00
10~20	0	0.00
20~35	0	0.00
35~60	0	0.00
≥60	0	0.00

17 小绿刺

xiǎo lǜ cì | Urophyllous Caper

Capparis urophylla F. Chun

白花菜科 Capparidaceae

代码（SpCode）= CAPURO

个体数（Individual number/15 hm^2）= 53

最大胸径（Max DBH）= 4.4 cm

重要值排序（Importance value rank）= 105

小乔木或灌木，高 2~7 m。树皮黑色，茎上刺粗壮，长达 5 mm，基部膨大，直或微外弯。叶顶端具延长尾，尾长 1~2.5 cm。花单出腋生或 2~3 多排成一短纵列腋上生。果球形，成熟后橘红色。花期 3~6 月，果期 8~12 月。

Small trees or shrubs, 2–7 m tall. Bark black, stipular spines on stems, ca. 5 mm, stout, straight or slightly recurved, base inflated. Leaf apex attenuate into a long tail, 1–2.5 cm long. Inflorescences uperaxillary rows, 1–3-flowered; Fruit globose, orangish red when maturity. Fl. Mar.–Jun, fr. Aug.–Dec..

花枝 Flowering branches

摄影：黄俞淞 Photo by：Huang Yusong

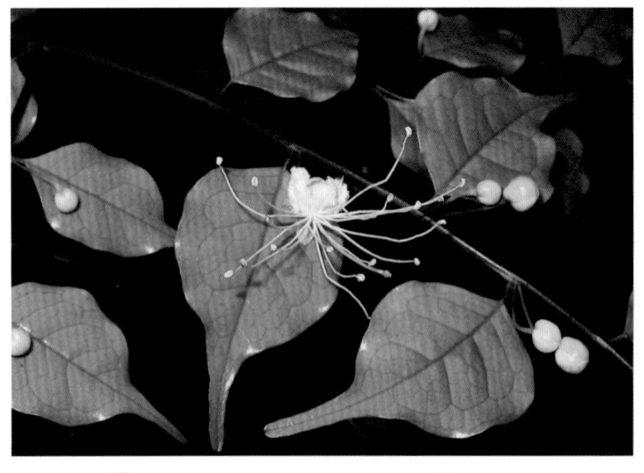

花 Flower

摄影：黄俞淞 Photo by：Huang Yusong

果枝 Fruiting branches

摄影：黄俞淞 Photo by：Huang Yusong

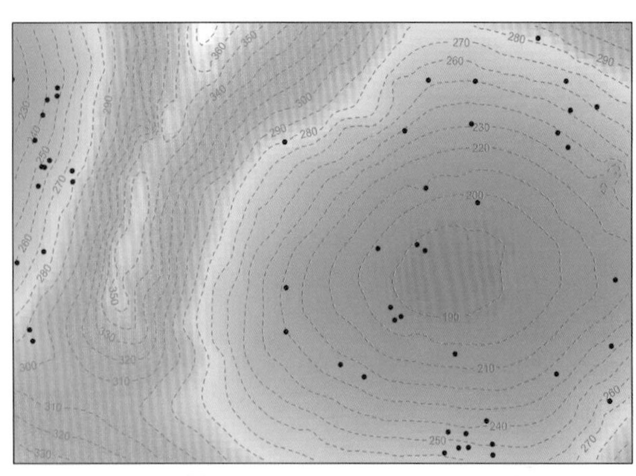

• 1~5 cm DBH + 5~20 cm DBH ○ ≥20 cm DBH

个体分布图 Distribution of individuals

径级分布表 DBH class

胸径区间 (Diameter class) (cm)	个体数 (No. of individuals in the plot)	比例 (Proportion) (%)
1~2	40	75.47
2~5	13	24.53
5~10	0	0.00
10~20	0	0.00
20~35	0	0.00
35~60	0	0.00
≥60	0	0.00

18 三角车

sān jiǎo chē | Bengal Rinorea

Rinorea bengalensis (Wall.) Kuntze
堇菜科 Violaceae

代码（SpCode）= RINBEN
个体数（Individual number/15 hm^2）= 2178
最大胸径（Max DBH）= 19.2 cm
重要值排序（Importance value rank）= 19

落叶灌木或小乔木，高可达 11 m。幼枝有明显的叶痕。叶互生，叶背中脉与侧脉间常具簇毛。花白色，花梗长 1 cm。蒴果近球形，3 裂瓣。花期 4~5 月，果期 9 月。

Deciduous shrubs or small trees, up to 11 m. Young branches with conspicuous leaf scars. Leaves alternate,between midvein of leaf abaxial and lateral veins with tufted hair. Flowers white, pedicel to 1 cm. Capsule subglobose, 3-valved. Fl. Apr.–May, fr. Sep..

花枝 Flowering branch
摄影：黄俞淞 Photo by：Huang Yusong

果 Fruits
摄影：黄俞淞 Photo by：Huang Yusong

花 Flowers
摄影：黄俞淞 Photo by：Huang Yusong

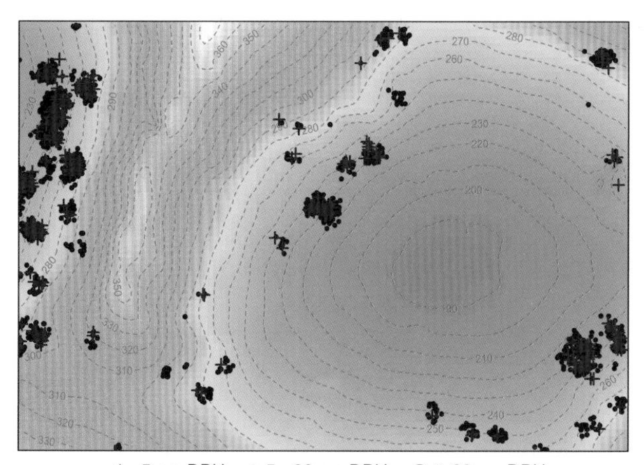

● 1~5 cm DBH ＋ 5~20 cm DBH ○ ≥20 cm DBH
个体分布图 Distribution of individuals

径级分布表 DBH class

胸径区间 (Diameter class) (cm)	个体数 (No. of individuals in the plot)	比例 (Proportion) (%)
1~2	871	39.99
2~5	985	45.22
5~10	296	13.59
10~20	26	1.19
20~35	0	0.00
35~60	0	0.00
≥60	0	0.00

19 米念芭 mǐ niàn bā | Ovoid Tirpitzia

Tirpitzia ovoidea Chun et How ex W. L. Sha
亚麻科 Linaceae

代码（SpCode）= TIROVO
个体数（Individual number/15 hm^2）= 416
最大胸径（Max DBH）= 14.9 cm
重要值排序（Importance value rank）= 59

常绿灌木或小乔木，高 0.5~4 m。树枝灰褐色，光滑。叶革质，叶缘全缘，顶端钝至稍凹陷。花瓣 5，白色，花柱 5 枚。蒴果卵状圆形。花期 5~10 月，果期 10~11 月。

Evergreen shrubs or small trees, 0.5–4 m tall. Branches brown to gray, glabrous. Leaf leathery, margin entire, apex obtuse to slightly concave. Petals 5, white, styles 5. Capsule ovoid-ellipsoid. Fl. May–Oct., fr. Oct.–Nov..

树干 Trunk
摄影：王斌 Photo by: Wang Bin

花枝 Flowering branch
摄影：黄俞淞 Photo by: Huang Yusong

果序 Infructescence
摄影：丁涛 Photo by: Ding Tao

• 1~5 cm DBH + 5~20 cm DBH ○ ≥20 cm DBH
个体分布图 Distribution of individuals

径级分布表 DBH class

胸径区间 (Diameter class) (cm)	个体数 (No. of individuals in the plot)	比例 (Proportion) (%)
1~2	95	22.84
2~5	215	51.68
5~10	95	22.84
10~20	11	2.64
20~35	0	0.00
35~60	0	0.00
≥60	0	0.00

20 尾叶紫薇

wěi yè zǐ wēi | Tail-leaf Crape-myrtle

Lagerstroemia caudata Chun et F. C. How ex S. Lee et L. F. Lau
千屈菜科 Lythraceae

代码（SpCode）= LAGCAU
个体数（Individual number/15 hm^2）= 60
最大胸径（Max DBH）= 28.1 cm
重要值排序（Importance value rank）= 90

落叶乔木，高 18~30 m。树枝光滑无毛。叶互生，稀近对生，叶片纸质或微革质，宽椭圆形，稀卵状椭圆形，叶背无毛或近叶脉被毛，侧脉 5~7 对。圆锥花序生于主枝及分枝顶端，长 3.5~8 cm。蒴果距圆状球形，5~6 裂。花期 4~5 月，果期 7~10 月。

Deciduous trees, 18–30 m tall. Branchlets glabrous. Leaves alternate, rarely subopposite, leaf blade papery to slightly leathery, broadly elliptic, rarely ovate-elliptic, abaxially glabrous or pubescent on veins, lateral veins 5–7 pairs. Panicles 3.5–8 cm, grow at the apex of branchs or blanchlet. Capsules oblong-globose, 5–6 valved. Fl. Apr.–May, fr. Jul.–Oct..

树干 Trunk
摄影：王斌 Photo by：Wang Bin

枝叶 Branch and leaves
摄影：黄俞淞 Photo by：Huang Yusong

叶 Leaves
摄影：王斌 Photo by：Wang Bin

● 1~5 cm DBH　+ 5~20 cm DBH　○ ≥20 cm DBH
个体分布图 Distribution of individuals

径级分布表 DBH class

胸径区间 (Diameter class) (cm)	个体数 (No. of individuals in the plot)	比例 (Proportion) (%)
1~2	6	10.00
2~5	21	35.00
5~10	18	30.00
10~20	13	21.67
20~35	2	3.33
35~60	0	0.00
≥60	0	0.00

21 秀丽海桐

xiù lì hǎi tóng | Beautiful Pittosporum

Pittosporum pulchrum Gagnep.
海桐花科 Pittosporaceae

代码（SpCode）= PITPUL
个体数（Individual number/15 hm^2）= 183
最大胸径（Max DBH）= 8.5 cm
重要值排序（Importance value rank）= 82

常绿灌木，高 3 m。老枝灰褐色，皮孔稀疏。叶多片（约）20 生于嫩枝顶，厚革质，倒卵形至倒披针形，先端圆，有时微凹入，叶缘通常会反卷。伞房花序顶生。蒴果圆球形，果爿薄木质，2 瓣裂开，种子约 15 个。花期 2~4 月，果期 3~10 月。

Evergreen shrubs, 3 m tall. Old branchlets gray-brown, sparsely lenticellate. Leaves usually ca. 20, clustered at branchlet apex, thickly leathery, obovate or oblanceolate, apex rounded, sometimes emarginated, margin usually reflexed. Corymb terminal; Capsule globose, dehiscing by 2 valves, seeds ca.15. Fl. Feb.–Apr., fr. Mar.–Oct..

果枝 Fruiting branch
摄影：黄俞淞 Photo by：Huang Yusong

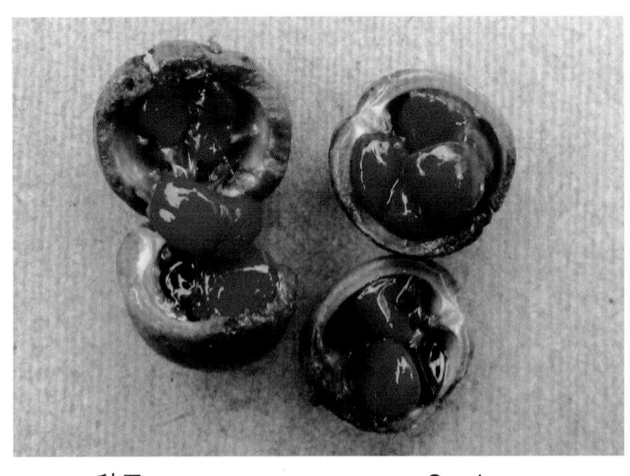

种子 Seeds
摄影：黄俞淞 Photo by：Huang Yusong

花序 Inflorescence
摄影：刘晟源 Photo by：Liu Shengyuan

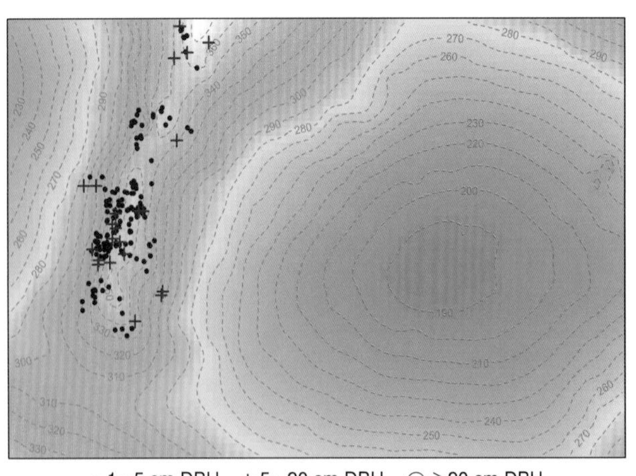

• 1~5 cm DBH　+ 5~20 cm DBH　○ ≥20 cm DBH
个体分布图 Distribution of individuals

径级分布表 DBH class

胸径区间 (Diameter class) (cm)	个体数 (No. of individuals in the plot)	比例 (Proportion) (%)
1~2	62	33.88
2~5	99	54.10
5~10	22	12.02
10~20	0	0.00
20~35	0	0.00
35~60	0	0.00
≥60	0	0.00

22 山桂花 shān guì huā | Common Bennettiodendron

Bennettiodendron leprosipes (Clos) Merr.
大风子科 Flacourtiaceae

代码（SpCode）= BENLEP
个体数（Individual number/15 hm^2）= 289
最大胸径（Max DBH）= 21.4 cm
重要值排序（Importance value rank）= 73

常绿灌木或小乔木。小枝有灰黄色短毛。叶窄或宽椭圆形，长椭圆形或倒卵形，边缘有疏离的钝锯齿，叶柄长 3~13 mm，有毛。圆锥花序顶生，初时密被短柔毛。浆果直径 6~9 mm。花期 3~4 月，果期 5~10 月。

Evergreen shrubs or small trees, branchlets densely gray-brown puberulous. Leaf blade mostly narrowly to broadly elliptic, elliptic-oblong, or obovate, margin sparsely obtusely serrate, petiole 3–13 mm, brown puberulous. Inflorescence terminal, paniculate, initially densely brown puberulous. Berry red when maturity, 6–9 mm in diam.. Fl. Mar.–Apr., fr. May–Nov..

幼苗 Seedling
摄影：黄俞淞 Photo by：Huang Yusong

花序 Inflorescence
摄影：黄俞淞 Photo by：Huang Yusong

果序 Infructescence
摄影：黄俞淞 Photo by：Huang Yusong

● 1~5 cm DBH + 5~20 cm DBH ○ ≥20 cm DBH
个体分布图 Distribution of individuals

径级分布表 DBH class

胸径区间 (Diameter class) (cm)	个体数 (No. of individuals in the plot)	比例 (Proportion) (%)
1~2	207	71.63
2~5	81	28.03
5~10	0	0.00
10~20	0	0.00
20~35	1	0.35
35~60	0	0.00
≥60	0	0.00

23 海南大风子

hǎi nán dà fēng zǐ | Hainan Chaulmoogra Tree

Hydnocarpus hainanensis (Merr.) Sleum.
大风子科 Flacourtiaceae

代码（SpCode）= HYDHAI
个体数（Individual number/15 hm^2）= 2260
最大胸径（Max DBH）= 32.2 cm
重要值排序（Importance value rank）= 7

常绿乔木，高 6~12 米。树皮灰褐色。叶近全缘或具疏离的不规则锯齿，两面无毛。花 15~20 朵，呈总状花序，腋生或顶生。浆果球形，表面密生棕褐色茸毛，果皮革质。花期 4~5 月，果期 6~8 月。

Evergreen trees, 6–12 m tall. Bark gray-brown. Leaves entire, or sparsely irregularly repand, both surfaces glabrous. Inflorescence axillary or terminal, 15–20 flowers in much condensed shortly pedunculate cymes. Berry globose, densely pale to dark brown or yellowish tomentose, pericarp leathery, Fl. Apr.–May, fr. Jun.–Aug..

树干 Trunk
摄影：王斌 Photo by：Wang Bin

果 Fruit
摄影：黄俞淞 Photo by：Huang Yusong

花枝 Flowering branch
摄影：黄俞淞 Photo by：Huang Yusong

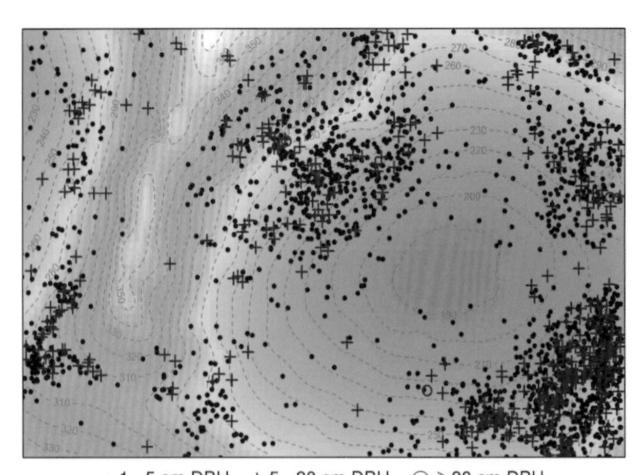

• 1~5 cm DBH + 5~20 cm DBH ○ ≥20 cm DBH
个体分布图 Distribution of individuals

径级分布表 DBH class

胸径区间 (Diameter class) (cm)	个体数 (No. of individuals in the plot)	比例 (Proportion) (%)
1~2	924	40.88
2~5	906	40.09
5~10	349	15.44
10~20	76	3.36
20~35	5	0.22
35~60	0	0.00
≥60	0	0.00

24 长叶柞木 cháng yè zuò mù | Long-leaf Xylosma

Xylosma longifolia Clos
大风子科 Flacourtiaceae

代码（SpCode）= XYLLON
个体数（Individual number/15 hm^2）= 45
最大胸径（Max DBH）= 12.6 cm
重要值排序（Importance value rank）= 114

常绿灌木或小乔木，高达 7 m。小枝有枝刺。叶矩圆形、矩圆状披针形至披针形，边缘有粗锯齿。短的总状花序或圆锥花序单生或簇生于叶腋内。浆果球形，成熟时黑色。花期 4~5 月，果期 6~10 月。

Evergreen shrubs or small trees, up to 7 m tall. Branchlcts spiny. Leaf blade oblong-elliptic, oblong-lanceolate, or lanceolate, margin serrate. Inflorescence of short racemes or reduced panides borne singly or in condensed clusters in leaf axils. Berry globose, black when maturity. Fl. Apr.–May, fr. Jun.–Oct..

树干 Trunks
摄影：王斌 Photo by: Wang Bin

枝叶 Branch and leaves
摄影：黄俞淞 Photo by: Huang Yusong

嫩叶 New Leaves
摄影：黄俞淞 Photo by: Huang Yusong

• 1~5 cm DBH + 5~20 cm DBH ○ ≥20 cm DBH
个体分布图 Distribution of individuals

径级分布表 DBH class

胸径区间 (Diameter class) (cm)	个体数 (No. of individuals in the plot)	比例 (Proportion) (%)
1~2	17	37.78
2~5	15	33.33
5~10	10	22.22
10~20	3	6.67
20~35	0	0.00
35~60	0	0.00
≥60	0	0.00

25 柳叶天料木 liŭ yè tiān liào mù | Willow-leaf Homalium

Homalium sabiifolium F. C. How et W. C. Ko
天料木科 Samydceae

代码（SpCode）= HOMSAB
个体数（Individual number/15 hm^2）= 56
最大胸径（Max DBH）= 7.0 cm
重要值排序（Importance value rank）= 121

常绿灌木，高 2~4 m。树皮棕黑色，嫩枝被柔毛。托叶线形，早落；叶柄短，长约 2 mm，被柔毛；叶片窄椭圆形，窄长椭圆形，革质，两面通常光滑无毛，边缘具浅钝齿。总状花序顶生或腋生，花多数，在花序轴上单生或双生。花期 10 月至翌年 2 月，果期 3~11 月。

Evergreen shrubs, 2–4 m tall. Bark gray-brown, branchlets pubescent when young. Stipules linear, early caducous; petiole short, ca. 2 mm, pubescent; leaf blade narrowly elliptic, rarely narrowly elliptic-oblong, leathery, glabrous, margin shallowly obtusely serrate, Inflorescence terminal or axillary, racemose. Flowers numerous, inserted along rachis singly or in pairs. Fl. Oct.–Feb., fr. Mar.–Nov..

花枝 Flowering branch
摄影：黄俞淞 Photo by：Huang Yusong

叶 Leaf
摄影：黄俞淞 Photo by：Huang Yusong

花序 Inflorescence
摄影：黄俞淞 Photo by：Huang Yusong

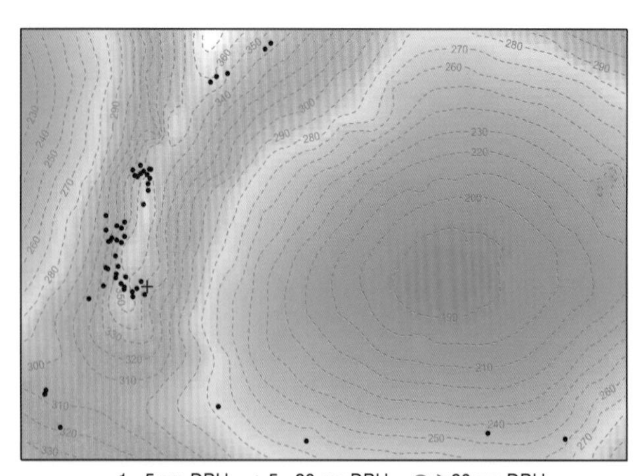

• 1~5 cm DBH + 5~20 cm DBH ○ ≥20 cm DBH
个体分布图 Distribution of individuals

径级分布表 DBH class

胸径区间 (Diameter class) (cm)	个体数 (No. of individuals in the plot)	比例 (Proportion) (%)
1~2	27	48.21
2~5	28	50.00
5~10	1	1.79
10~20	0	0.00
20~35	0	0.00
35~60	0	0.00
≥60	0	0.00

26 淡黄金花茶 dàn huáng jīn huā chá | Bright-yellow Camellia

Camellia flavida H. T. Chang
山茶科 Theaceae

代码（SpCode）= CAMFLA
个体数（Individual number/15 hm²）= 78
最大胸径（Max DBH）= 13.2 cm
重要值排序（Importance value rank）= 112

常绿灌木。嫩枝无毛，当年生小枝略带紫红色。叶薄革质，基部楔形，或近圆形，两面无毛，边缘有细齿。花顶生或腋生，淡黄色。蒴果球形或扁球形，种子被红棕色柔毛。花期 11 月至翌年 1 月，果期 9 月至 10 月。

Evergreen shrubs, young branches glabrous, current year branchlets purplish red. Leaves thinly leathery, base cuneate to obtuse, both surfaces glabrous, margin serrulate. Flowers terminal or axillary, faint yellow. Capsule globose or oblate, seeds with reddish brown villous. Fl. Nov.–Jan. of next year, fr. Sep.–Oct..

嫩叶 New Leaves
摄影：丁涛 Photo by：Ding Tao

花 Flower
摄影：丁涛 Photo by：Ding Tao

果 Fruits
摄影：黄俞淞 Photo by：Huang Yusong

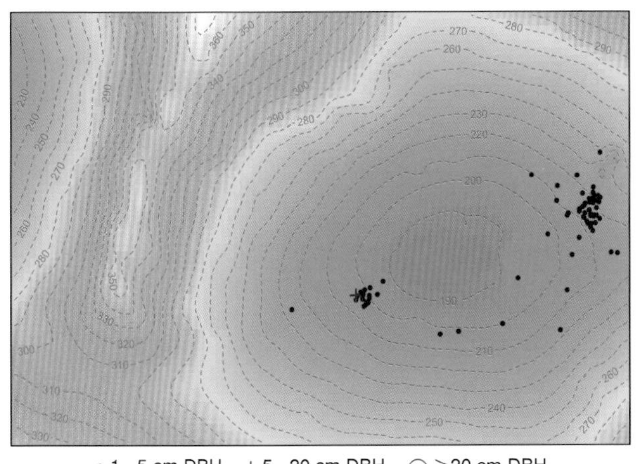

• 1~5 cm DBH + 5~20 cm DBH ○ ≥20 cm DBH
个体分布图 Distribution of individuals

径级分布表 DBH class

胸径区间 (Diameter class) (cm)	个体数 (No. of individuals in the plot)	比例 (Proportion) (%)
1~2	48	61.54
2~5	29	37.18
5~10	0	0.00
10~20	1	1.28
20~35	0	0.00
35~60	0	0.00
≥60	0	0.00

27 水东哥

shuǐ dōng gē | Common Saurauia

Saurauia tristyla DC.
水东哥科 Saurauiaceae

代码（SpCode） = SAUTRI
个体数（Individual number/15 hm^2） = 22
最大胸径（Max DBH） = 24.4 cm
重要值排序（Importance value rank） = 124

落叶灌木或小乔木。小枝无毛或被绒毛，被爪甲状鳞片或钻状刺毛。叶片倒卵或宽倒卵椭圆形，纸质或薄革质，两面中、侧脉具钻状刺毛或爪甲状鳞片。花序聚伞式，1~4 枚簇生于叶腋，被毛和鳞片。果球形，白色，绿色或淡黄色。花期 5~7 月，果期 8~12 月。

Deciduous shrubs or small trees. Branchlets tomentose, with unguiculate scales or subulate hairs. Leaf blade obovate to elliptic-obovate, papery or thinly leathery, both surfaces sparsely pubescent with appressed setose to ynguiculate hairs on midvein and lateral veins. Inflorescences 1–4-fascicled, axillary, hairy and scaly. Fruit green to white to pale yellow, globose. Fl. Mar.–Jul., fr. Aug.–Dec..

树干 Trunk
摄影：黄俞淞 Photo by：Huang Yusong

花序 Inflorescence
摄影：黄俞淞 Photo by：Huang Yusong

枝叶 Branch and leaves
摄影：黄俞淞 Photo by：Huang Yusong

• 1~5 cm DBH + 5~20 cm DBH ○ ≥20 cm DBH
个体分布图 Distribution of individuals

径级分布表 DBH class

胸径区间 (Diameter class) (cm)	个体数 (No. of individuals in the plot)	比例 (Proportion) (%)
1~2	2	9.09
2~5	1	4.55
5~10	3	13.64
10~20	15	68.18
20~35	1	4.55
35~60	0	0.00
≥60	0	0.00

28 子楝树

zǐ liàn shù | Slender Decaspermum

Decaspermum gracilentum (Hance) Merr. et Perry
桃金娘科 Myrtaceae

代码（SpCode）= DECGAR
个体数（Individual number/15 hm^2）= 624
最大胸径（Max DBH）= 35.6 cm
重要值排序（Importance value rank）= 33

灌木至小乔木。嫩枝常四棱形或具狭翅。叶片纸质或薄革质，椭圆形或长圆状披针形，基部楔形。花单生或数朵组成腋生的聚伞花序，白色或粉红色，花萼 3 或 4 枚。浆果直径 2~4 mm，有柔毛。花期 3~5 月，果期 9~10 月。

Shrubs or small trees. Branchlets often 4-angled or narrowly 4-winged. Leaf blade papery or slightly leathery, elliptic to oblong lanceolate, base cuneate. Flowers solitary, or several flowers of cymes in axil, white or pink, sepals 3–4. Fruit 2–4 mm in diam., sparsely pubescent. Fl. Mar.–May, fr. Sep.–Oct..

花枝　Flowering branch
摄影：黄俞淞　Photo by：Huang Yusong

果枝　Fruiting branch
摄影：黄俞淞　Photo by：Huang Yusong

花　Flower
摄影：黄俞淞　Photo by：Huang Yusong

• 1~5 cm DBH　+ 5~20 cm DBH　○ ≥20 cm DBH
个体分布图　Distribution of individuals

径级分布表　DBH class

胸径区间 (Diameter class) (cm)	个体数 (No. of individuals in the plot)	比例 (Proportion) (%)
1~2	176	28.21
2~5	311	49.84
5~10	117	18.75
10~20	19	3.04
20~35	0	0.00
35~60	1	0.16
≥60	0	0.00

29 密脉蒲桃

mì mài pú táo | Chun's Syzygium

Syzygium chunianum Merr. et Perry
桃金娘科 Myrtaceae

代码（SpCode）= SYZCHU
个体数（Individual number/15 hm^2）= 12
最大胸径（Max DBH）= 10.9 cm
重要值排序（Importance value rank）= 151

常绿乔木，高可达 22 m。嫩枝干后灰色。叶片薄革质椭圆形或倒卵状椭圆形，两面均有细小腺点，先端宽而急渐尖，尖头长 1~1.5 cm，侧脉多而密，近于水平缓斜向边缘。圆锥花序顶生或近顶生。果实球形。花期 6~7 月，果期 8~12 月。

Evergreen trees, to 22 m tall. Young branchlets gray when dry. Leaf blade thinly leathery, elliptic to obovate-elliptic, both surfaces with small glands, apex broadly and abruptly acuminate and with a 1–1.5 cm acumen, secondary veins numerous, less than 1 mm apart, and gradually extending into margin. Inflorescences teminal or subteminal, paniculate cymes. Fruit globose. Fl. Jun.–Jul., fr. Aug.–Dec..

树干 Trunk
摄影：王斌 Photo by：Wang Bin

枝叶 Branch and leaves
摄影：王斌 Photo by：Wang Bin

叶背 Leaf back
摄影：王斌 Photo by：Wang Bin

• 1~5 cm DBH　+ 5~20 cm DBH　○ ≥20 cm DBH

个体分布图 Distribution of individuals

径级分布表 DBH class

胸径区间 (Diameter class) (cm)	个体数 (No. of individuals in the plot)	比例 (Proportion) (%)
1~2	2	16.67
2~5	4	33.33
5~10	5	41.67
10~20	1	8.33
20~35	0	0.00
35~60	0	0.00
≥60	0	0.00

30 海南蒲桃

hǎi nán pú táo | Hainan Syzygium

Syzygium hainanense Chang et Miau
桃金娘科 Myrtaceae

代码（SpCode）= SYZHAI
个体数（Individual number/15 hm^2）= 2
最大胸径（Max DBH）= 6.2 cm
重要值排序（Importance value rank）= 192

常绿乔木，高 6~20 m。嫩枝干后灰白色。叶片革质，先端圆或钝，具一尖头，侧脉多而密，脉间相隔 1~2 mm。圆锥花序腋生或生于花枝上，长可达 11 cm。果实由红色变黑色。花期 2~3 月或 4~5 月，果期 6~9 月。

Evergreen trees, 6–20 m tall. Young branchlets grayish white. Leaf blade leathery, apex rounded or obtuse, and with a short cusp, secondary veins numerous, 1–2 mm apart. Inflorescences axillary on flowering branches, paniculate cymes, to 11 cm long. Fruit red to black. Fl. Feb.–Mar., or Apr.–May, fr. Jun.–Sep..

枝叶 Branch and leaves
摄影：黄俞淞 Photo by：Huang Yusong

花序 Inflorescence
摄影：郑锡荣 Photo by：Zheng Xirong

叶 Leaves
摄影：黄俞淞 Photo by：Huang Yusong

• 1~5 cm DBH + 5~20 cm DBH ○ ≥20 cm DBH
个体分布图 Distribution of individuals

径级分布表 DBH class

胸径区间 (Diameter class) (cm)	个体数 (No. of individuals in the plot)	比例 (Proportion) (%)
1~2	0	0.00
2~5	1	50.00
5~10	1	50.00
10~20	0	0.00
20~35	0	0.00
35~60	0	0.00
≥60	0	0.00

31 细叶谷木

xì yè gǔ mù | Little-leaf Memecylon

Memecylon scutellatum (Lour.) Hook. et Arn.
野牡丹科 Melastomataceae

代码（SpCode）= MEMSCU
个体数（Individual number/15 hm^2）= 1221
最大胸径（Max DBH）= 74.8 cm
重要值排序（Importance value rank）= 10

常绿灌木或小乔木，高 1.5~4 m。树皮灰色，小枝四棱形。叶革质，顶端钝，圆形或微凹，叶压干后两面粗糙，密布小突起。聚伞花序腋生。浆果状核果球形，密布小疣状突起。花期 3~8 月，果期 11 月至翌年 3 月。

Evergreen shrubs or small trees, 1.5–4 m tall. Bark gray, branches 4-sided. Leaf blade leathery, apex obtuse, rounded or retuse, both surfaces densely small tuberculate and scabrous when dry. Cyme axillary. Fruit a baccate drupe, densely tuberculate. Fl. Mar.–Aug., fr. Nov.–Mar. of next year.

树干 Trunk
摄影：王斌 Photo by: Wang Bin

花 Flowers
摄影：刘晟源 Photo by: Liu Shengyuan

果枝 Fruiting branches
摄影：黄俞淞 Photo by: Huang Yusong

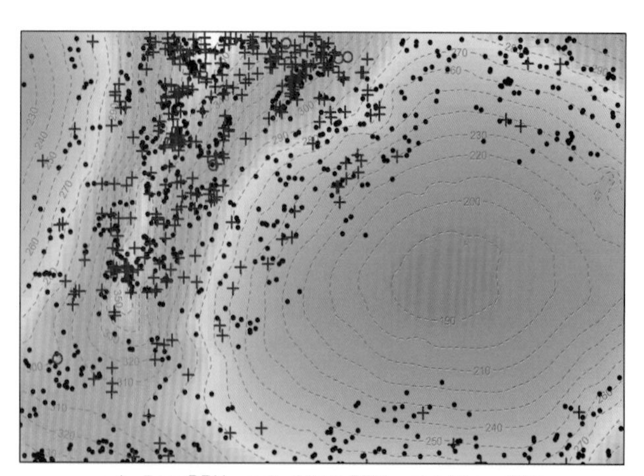

• 1~5 cm DBH + 5~20 cm DBH ○ ≥20 cm DBH
个体分布图 Distribution of individuals

径级分布表 DBH class

胸径区间 (Diameter class) (cm)	个体数 (No. of individuals in the plot)	比例 (Proportion) (%)
1~2	325	26.62
2~5	543	44.47
5~10	244	19.98
10~20	103	8.44
20~35	5	0.41
35~60	0	0.00
≥60	1	0.08

32 红芽木

hóng yá mù | Kudin Cratoxylum

Cratoxylum formosum subsp. *pruniflorum* (Kurz) Gogelein
金丝桃科 Hypericaceae

代码（SpCode）= CRAFOR
个体数（Individual number/15 hm^2）= 74
最大胸径（Max DBH）= 34.3 cm
重要值排序（Importance value rank）= 84

落叶乔木，高达 20 m。幼枝、叶和叶柄有灰褐色柔毛。叶椭圆状披针形，腹面疏生短毛，背面密被灰白色柔毛。花序腋生。蒴果棕黑色，椭圆形。花期 3~5 月，果期 5 月后。

Deciduous trees, up to 20 m. Young branches, leaves and petiole with gray-brown pubescence. Leaf blade elliptic-lanceolate, adaxially with sparsely short hairs, abaxially densely gray-white pubescence. Inflorescence axillary. Capsule dark brown, oblong. Fl. Mar.–May, fr. after May.

树干 Trunk
摄影：王斌 Photo by: Wang Bin

枝叶 Branch and leaves
摄影：黄俞淞 Photo by: Huang Yusong

花枝 Flowering branch
摄影：丁涛 Photo by: Ding Tao

• 1~5 cm DBH + 5~20 cm DBH ○ ≥20 cm DBH
个体分布图 Distribution of individuals

径级分布表 DBH class

胸径区间 (Diameter class) (cm)	个体数 (No. of individuals in the plot)	比例 (Proportion) (%)
1~2	32	43.24
2~5	19	25.68
5~10	9	12.16
10~20	7	9.46
20~35	7	9.46
35~60	0	0.00
≥60	0	0.00

33 大苞藤黄

dà bāo téng huáng | Large-Bract Garcinia

Garcinia bracteata C. Y. Wu ex Y. H. Li
藤黄科 Guttiferae

代码（SpCode）= GARBRA
个体数（Individual number/15 hm^2）= 3
最大胸径（Max DBH）= 1.9 cm
重要值排序（Importance value rank）= 185

常绿乔木，高约 8 m。树皮红褐色，薄片状脱落。叶卵形或椭圆形至长圆形，叶背侧脉明显，密集，20~30 对，叶干后变黄色。伞形花序腋生，基部有 2 枚叶状苞片。果卵形，先端歪斜。花期 4~5 月，果期 11~12 月。

Evergreen trees, ca. 8 m tall. Bark reddish-brown, lamellar abscission. Leaves ovate or oval to long circle, abaxial veins conspicuous, densely, 20–30 pairs, yellow when dry. Umbel axillary, basal with 2 leaf-shaped bracts. Fruit ovate, apex askew. Fl. Apr.–May, fr. Nov.–Dec..

枝叶 Branch and leaves
摄影：黄俞淞 Photo by：Huang Yusong

花枝 Flowering branch
摄影：刘晟源 Photo by：Liu Shengyuan

果 Fruits
摄影：黄俞淞 Photo by：Huang Yusong

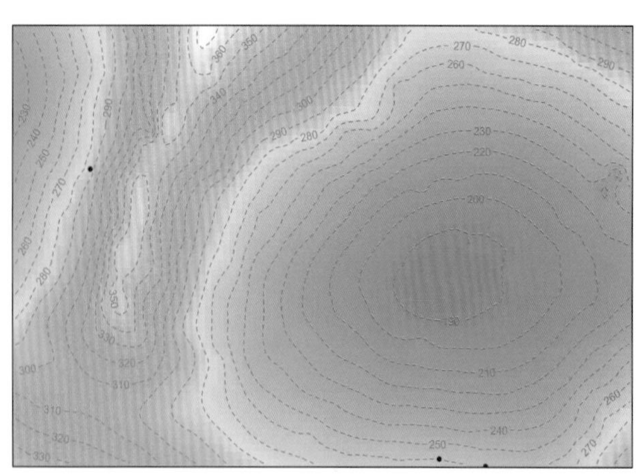

• 1~5 cm DBH　+ 5~20 cm DBH　○ ≥20 cm DBH
个体分布图 Distribution of individuals

径级分布表 DBH class

胸径区间 (Diameter class) (cm)	个体数 (No. of individuals in the plot)	比例 (Proportion) (%)
1~2	3	100.00
2~5	0	0.00
5~10	0	0.00
10~20	0	0.00
20~35	0	0.00
35~60	0	0.00
≥60	0	0.00

34 金丝李

jīn sī lǐ | Few-nerve Garcinia

Garcinia paucinervis Chun ex F. C. How
藤黄科 Guttiferae

代码（SpCode） = GARPAU
个体数（Individual number/15 hm^2） = 1684
最大胸径（Max DBH） = 41.4 cm
重要值排序（Importance value rank） = 8

常绿乔木，高 3~15 m。树皮灰黑色，具白斑块。叶片嫩时紫红色，侧脉 5~8 对，中脉在下面凸起。雌雄同株。果成熟时椭圆形或卵珠状椭圆形，黄色略带红色，基部萼片宿存, 种子 2 颗。花期 6~7 月，果期 11~12 月。

Evergreen trees, 3–15 m tall. Bark gray-black, with white spotted. Leaf blade purplered when young, lateral veins 5–8, midrib conspicuous raised on leaf abaxial.Plant monoecious, mature fruit ellipsoid or ovoid-ellipsoid, yellow with slightly red, basal sepals persistent, seeds 2. Fl. Jun.–Jul., fr. Nov.–Dec..

树干 Trunk
摄影：王斌 Photo by：Wang Bin

花枝 Flowering branch
摄影：黄俞淞 Photo by：Huang Yusong

叶 Leaves
摄影：王斌 Photo by：Wang Bin

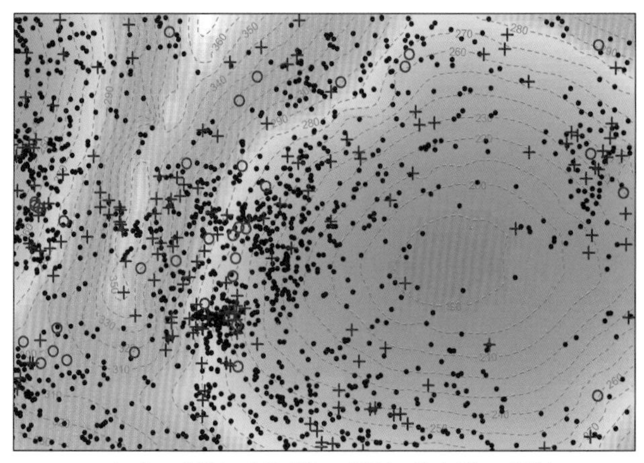

● 1~5 cm DBH + 5~20 cm DBH ○ ≥20 cm DBH
个体分布图 Distribution of individuals

径级分布表 DBH class

胸径区间 (Diameter class) (cm)	个体数 (No. of individuals in the plot)	比例 (Proportion) (%)
1~2	766	45.49
2~5	713	42.34
5~10	125	7.42
10~20	46	2.73
20~35	29	1.72
35~60	5	0.30
≥60	0	0.00

35 海南椴 hǎi nán duàn | Hairy-seeded Hainania

Diplodiscus trichosperma (Merr.) Y. Tang, M. G. Gilbert et Dorr
椴树科 Tiliaceae

代码（SpCode）= DIPTRI
个体数（Individual number/15 hm^2）= 1126
最大胸径（Max DBH）= 45.5 cm
重要值排序（Importance value rank）= 5

落叶乔木，高达 15 m。树皮灰白色。叶基部微心形或截形，腹面无毛，背面密被紧贴的灰黄色星状短绒毛，基出脉 5~7 条。圆锥花序顶生，长达 26 cm。蒴果倒卵形，有 4~5 棱，种子椭圆形，密被柔毛。花期秋季，果期冬季。

Deciduous trees, up to 15 m. Bark gray-white. Leaf base subcordate or truncate, adaxially glabrous, abaxially densely appressed gray-yellow stellate puberulent, basal veins 5–7. Panicles to 26 cm, terminal. Capsule obovate, with 4–5 longitudinal ridges, seeds oval, densely yellowish stellate puberulent. Fl. Autumn, fr. Winter.

树干 Trunk
摄影：王斌 Photo by: Wang Bin

果枝 Fruiting branch
摄影：黄俞淞 Photo by: Huang Yusong

枝叶 Branch and leaves
摄影：刘晟源 Photo by: Liu Shengyuan

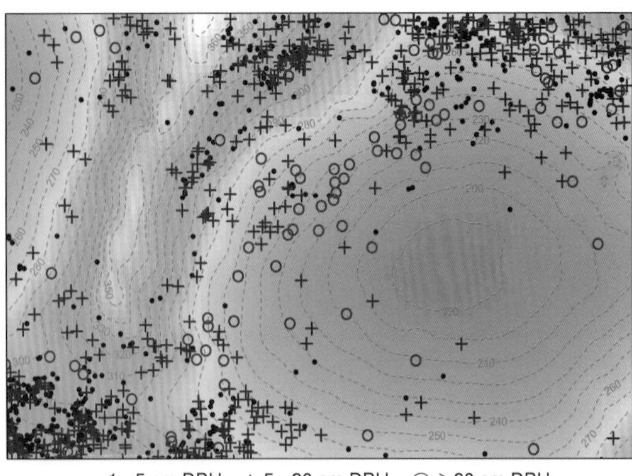

个体分布图 Distribution of individuals

径级分布表 DBH class

胸径区间 (Diameter class) (cm)	个体数 (No. of individuals in the plot)	比例 (Proportion) (%)
1~2	192	17.05
2~5	341	30.28
5~10	272	24.16
10~20	229	20.34
20~35	84	7.46
35~60	8	0.71
≥60	0	0.00

36 蚬木 xiàn mù | Hsienmu

Excentrodendron tonkinense (A. Chev.) H. T. Chang et R. H. Miao
椴树科 Tiliaceae

代码（SpCode） = EXCTON
个体数（Individual number/15 hm^2） = 1502
最大胸径（Max DBH） = 84.1 cm
重要值排序（Importance value rank） = 4

常绿乔木，高可达 30 m 以上。叶革质，卵圆形或椭圆形，叶基部圆形，背面除脉腋有毛丛外秃净无毛，基出脉 3 条。圆锥花序。翅果椭圆形，具 5 条薄翅。花期 3 月，果期 5~6 月。

Evergreen trees, up to 30 m tall. Leaf blade leathery, orbicular-ovate or elliptic-ovate, base rounded, glabrous except abaxially yellow-brown fascided hairy in vein axils, basal veins 3. Inflorescences paniculate. Capsule ellipsoid，with 5-winged. Fl. Mar., fr. May–Jun..

树干 Trunk
摄影：王斌 Photo by：Wang Bin

果枝 Fruiting branch
摄影：黄俞淞 Photo by：Huang Yusong

枝叶 Branch and leaves
摄影：王斌 Photo by：Wang Bin

• 1~5 cm DBH + 5~20 cm DBH ○ ⩾20 cm DBH
个体分布图 Distribution of individuals

径级分布表 DBH class

胸径区间 (Diameter class) (cm)	个体数 (No. of individuals in the plot)	比例 (Proportion) (%)
1~2	569	37.88
2~5	419	27.90
5~10	204	13.58
10~20	190	12.65
20~35	92	6.13
35~60	27	1.80
⩾60	1	0.07

37 黄麻叶扁担杆 huáng má yè biǎn dàn gǎn | Henry's Grewia

Grewia henryi Burret
椴树科 Tiliaceae

代码（SpCode）= GREHEN
个体数（Individual number/15 hm²）= 116
最大胸径（Max DBH）= 11.6 cm
重要值排序（Importance value rank）= 78

常绿灌木或小乔木，高 1~6 m。嫩枝被黄褐色星状粗毛。叶薄革质，叶面有稀疏星状短粗毛，叶背被黄绿色星状粗毛，叶柄被星状粗毛。聚伞花序 1~2 枝腋生，每枝有花 3~4 朵。核果 4 裂。花期 5~6 月。

Evergreen shrubs or small trees, 1–6 m tall. Branchlets yellow-brown coarsely stellate. Leaf blade thinly leathery, adaxially sparsely shortly stellate hairy, abaxially coarsely yellow-green stellate hairy, petiole coarsely stellate. Cymes 1 or 2 per leaf axil, 3–4-flowered. Drupe 4-lobed. Fl. May–Jun..

花枝 Flowering branch
摄影：刘晟源 Photo by：Liu Shengyuan

花 Flower
摄影：黄俞淞 Photo by：Huang Yusong

果枝 Fruiting branch
摄影：黄俞淞 Photo by：Huang Yusong

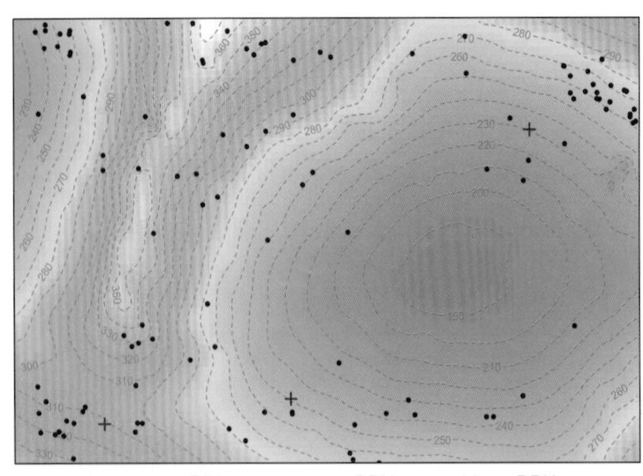

• 1~5 cm DBH + 5~20 cm DBH ○ ≥20 cm DBH
个体分布图 Distribution of individuals

径级分布表 DBH class

胸径区间 (Diameter class) (cm)	个体数 (No. of individuals in the plot)	比例 (Proportion) (%)
1~2	69	59.48
2~5	44	37.93
5~10	2	1.72
10~20	1	0.86
20~35	0	0.00
35~60	0	0.00
≥60	0	0.00

38 梧桐

wú tóng | Chinese Parasoltree

Firmiana simplex (L.) W. Wight
梧桐科 Sterculiaceae

代码（SpCode）= FIRSIM
个体数（Individual number/15 hm^2）= 10
最大胸径（Max DBH）= 15.4 cm
重要值排序（Importance value rank）= 160

落叶乔木，高达 16 m。树皮青绿色，平滑。叶心形，掌状 3~5 裂，基部心形，两面均无毛或略被短柔毛，基生脉 7 条。圆锥花序顶生，长约 20~50 cm，蓇葖果膜质，外面被短茸毛或几无毛. 花期 6 月。

Deciduous trees, up to 16 m tall. Bark greenish, smooth. Leaf blade cordate, palmately 3–5-lobed, base cordate, both surfaces glabrous or minutely puberulent, basal veins 7. Inflorescence paniculate, ca. 20–50 cm long. Follicle membranous, abaxially puberulent or nearly glabrous. Fl. Jun..

树干 Trunk
摄影：黄俞淞 Photo by：Huang Yusong

果枝 Fruiting branches
摄影：黄俞淞 Photo by：Huang Yusong

叶 Leaves
摄影：黄俞淞 Photo by：Huang Yusong

• 1~5 cm DBH + 5~20 cm DBH ○ ≥20 cm DBH
个体分布图 Distribution of individuals

径级分布表 DBH class

胸径区间 (Diameter class) (cm)	个体数 (No. of individuals in the plot)	比例 (Proportion) (%)
1~2	3	30.00
2~5	3	30.00
5~10	3	30.00
10~20	1	10.00
20~35	0	0.00
35~60	0	0.00
≥60	0	0.00

39 截裂翅子树 jié liè chì zǐ shù | Truncate-lobed Wingseedtree

Pterospermum truncatolobatum Gagnep.
梧桐科 Sterculiaceae

代码（SpCode）= PTETRU
个体数（Individual number/15 hm^2）= 1604
最大胸径（Max DBH）= 30.6 cm
重要值排序（Importance value rank）= 12

常绿乔木，高 16 m。树皮黑色，小枝嫩部分密被黄褐色星状绒毛。叶革质，椭圆状倒卵形，先端有 3~5 裂，叶背密被黄褐色和灰白色星状绒毛。花腋生，单生，几无梗。蒴果木质，有明显的 5 棱和 5 条深沟，外面密被黄褐色星状绒毛。花期 7~8 月，果期 10~11 月。

Evergreen trees, to 16 m tall. Bark black, branchlets densely yellow-brown stellate tomentose. Leaf blade leathery, oblong-obovate, apex 3–5-lobed, abaxially densely yellow-brown or gray-white tomentose. Flowers axillary, solitary, nearly sessile. Capsule woody, prominently 5-angular and 5-grooved, densely yellow-brow stellate tomentose. Fl. Jul.–Aug., fr, Oct.–Nov..

树干 Trunk
摄影：王斌 Photo by：Wang Bin

叶背 Leaf back
摄影：黄俞淞 Photo by：Huang Yusong

枝叶 Branch and leaves
摄影：黄俞淞 Photo by：Huang Yusong

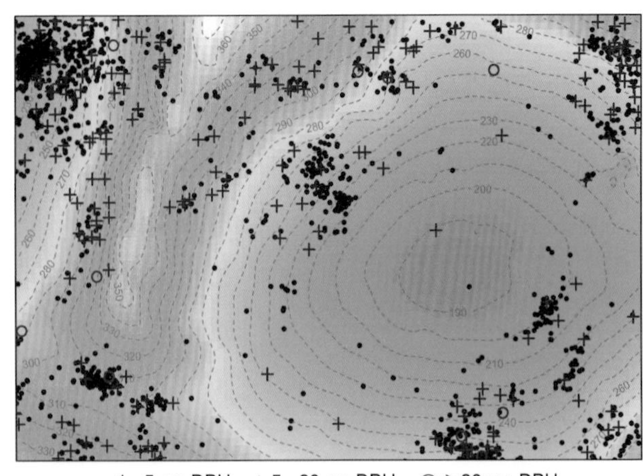

• 1~5 cm DBH + 5~20 cm DBH ○ ≥20 cm DBH
个体分布图 Distribution of individuals

径级分布表 DBH class

胸径区间 (Diameter class) (cm)	个体数 (No. of individuals in the plot)	比例 (Proportion) (%)
1~2	707	44.08
2~5	686	42.77
5~10	142	8.85
10~20	57	3.55
20~35	12	0.75
35~60	0	0.00
≥60	0	0.00

40 瑶山梭罗 yáo shān suō luó | Pale-leaf Reevesia

Reevesia glaucophylla Hsue
梧桐科 Sterculiaceae

代码（SpCode）= REEGLA
个体数（Individual number/15 hm^2）= 35
最大胸径（Max DBH）= 14.6 cm
重要值排序（Importance value rank）= 129

常绿乔木，高 8~16 m。树皮有纵裂纹。叶纸质或薄革质，基部钝、圆形或浅心形，上面沿主脉和侧脉被淡黄褐色星状短柔毛，下面密被灰白色星状短柔毛常有白霜。聚伞状伞房花序顶生，被淡黄褐色星状短柔毛。蒴果 5 棱，密被淡黄褐色星状短柔毛。花期 5~6 月。

Evergreen trees, 8–16 m tall. Bark exfoliating longitudinally. Leaf blade papery or thinly leathery, base obtuse, rounded or shallowly cordate, adaxially yellowish brown stellate puberulent along veins and midrib, abaxially densely gray stellate puberulent. Inflorescence cymose, terminal, yellowish brown stellate puberulent. Capsule 5-angled, densely yellowish brown stellate puberulent. Fl. May–Jun..

树干 Trunk
摄影：王斌 Photo by：Wang Bin

叶 Leaves
摄影：徐永福 Photo by：Xu yongfu

枝叶 Branch and leaves
摄影：徐永福 Photo by：Xu yongfu

• 1~5 cm DBH + 5~20 cm DBH ○ ≥20 cm DBH
个体分布图 Distribution of individuals

径级分布表 DBH class

胸径区间 (Diameter class) (cm)	个体数 (No. of individuals in the plot)	比例 (Proportion) (%)
1~2	11	31.43
2~5	12	34.29
5~10	9	25.71
10~20	3	8.57
20~35	0	0.00
35~60	0	0.00
≥60	0	0.00

41 粉苹婆

fěn píng pó | Well-flavored Sterculia

Sterculia euosma W. W. Sm.
梧桐科 Sterculiaceae

代码（SpCode）= STEEUO
个体数（Individual number/15 hm^2）= 70
最大胸径（Max DBH）= 16.1 cm
重要值排序（Importance value rank）= 106

常绿乔木。嫩枝密被淡黄褐色绒毛。叶革质，基部圆形或略为斜心形，有基出脉 5 条，叶面无毛，叶背密被淡黄褐色星状茸毛。总状花序聚生于小枝上部。蓇葖果熟时红色，顶端渐尖成喙状，外面密被星状短茸毛。花期 4~5 月，果期 7~8 月。

Evergreen trees. Young branchlets densely yellowish brown villous. Leaf blade leathery, base rounded or nearly obliquely cordae, basal veins 5, adaxially glabrous, abaxially densely yellowish brown stellate villous. Inflorescence racemose, clustered subapically on branchlets. Follicle red when mature, apex acuminate into beak, densely stellate villous. Fl. Apr.–May, fr. Jul.–Aug..

植株 Whole plant
摄影：刘晟源 Photo by：Liu Shengyuan

叶 Leaves
摄影：黄俞淞 Photo by：Huang Yusong

果 Fruit
摄影：黄俞淞 Photo by：Huang Yusong

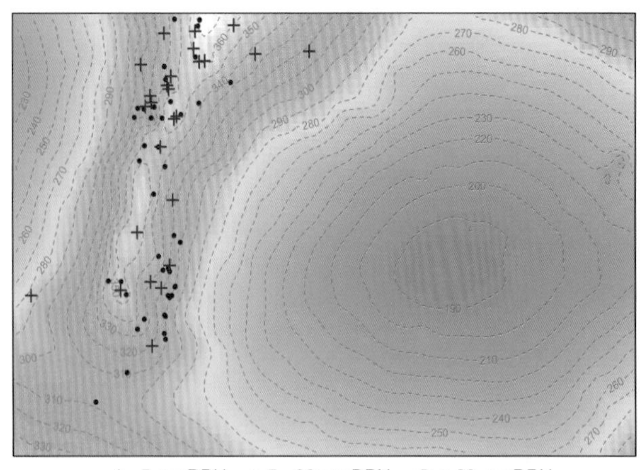

• 1~5 cm DBH + 5~20 cm DBH ○ ≥20 cm DBH
个体分布图 Distribution of individuals

径级分布表 DBH class

胸径区间 (Diameter class) (cm)	个体数 (No. of individuals in the plot)	比例 (Proportion) (%)
1~2	19	27.14
2~5	25	35.71
5~10	18	25.71
10~20	8	11.43
20~35	0	0.00
35~60	0	0.00
≥60	0	0.00

42 苹婆 píng pó | Thai Chestnut

Sterculia monosperma Vent.
梧桐科 Sterculiaceae

代码（SpCode）= STEMON
个体数（Individual number/15 hm^2）= 6328
最大胸径（Max DBH）= 44.0 cm
重要值排序（Importance value rank）= 2

常绿乔木。树皮褐黑色。单叶薄革质，长圆形或椭圆形。圆锥花序顶生或腋生，花萼初时乳白色，后转为淡红色，先端互相黏合，与钟状萼筒等长。骨葖果鲜红色，先端有喙。花期 4~5 月，少数植株 10~11 月间第二次开花, 果期 7~8 月。

Evergreen trees. Bark brown-black. Leave simple, thinly leathery, leaf blade ovate or oblong. Inflorescence paniculate, terminal or axillary, calyx cream-white, becoming reddish, apex acuminate, incurved and cohering apically, as long as calyx tube. Follicles red, apex beaked. Fl. Apr.–May, rarely reflorescence Oct.–Nov., fr. Jul.–Aug..

树干 Trunk
摄影：王斌 Photo by：Wang Bin

果枝 Fruiting branches
摄影：黄俞淞 Photo by：Huang Yusong

叶 Leaves
摄影：王斌 Photo by：Wang Bin

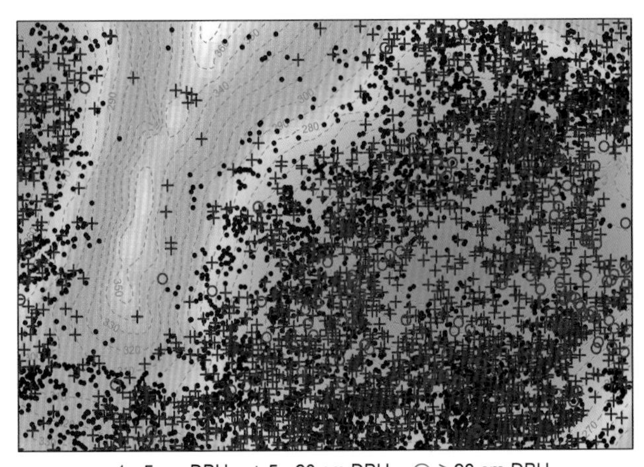

• 1~5 cm DBH + 5~20 cm DBH ○ ≥20 cm DBH
个体分布图 Distribution of individuals

径级分布表 DBH class

胸径区间 (Diameter class) (cm)	个体数 (No. of individuals in the plot)	比例 (Proportion) (%)
1~2	1696	26.80
2~5	2160	34.13
5~10	1391	21.98
10~20	918	14.51
20~35	158	2.50
35~60	5	0.08
≥60	0	0.00

43 木棉

mù mián | red silk cottontree

Bombax ceiba L.
木棉科 Bombacaceae

代码（SpCode）= BOMCEI
个体数（Individual number/15 hm^2）= 3
最大胸径（Max DBH）= 68.3 cm
重要值排序（Importance value rank）= 138

落叶大乔木，高可达 25 m。树皮灰白色，幼树的树干通常有圆锥状粗刺。掌状复叶，小叶 5~7 片，全缘，两面均无毛。花单生枝顶叶腋，花瓣肉质，通常红色，倒卵状长圆形，两面被星状柔毛。蒴果长圆形，密被灰白色长柔毛和星状柔毛。花期 3~4 月，果夏季成熟。

Deciduous trees, up to 25 m tall. Bark pallid, trunk usually has conical spine. Leaves palmate, leaflets 5–7, entire, glabrous on both face. Flowers solitary, temina, petal fleshy, usually red, obovate-oblong, both surfaces stellate puberulent. Capsule ellipsoid, densely gray-white villous and stellate puberulent. Fl. Mar.–Apr., fr. ripe in summer.

花枝 Flowering branches
摄影：刘晟源 Photo by：Liu Shengyuan

枝叶 Branch and leaves
摄影：黄俞淞 Photo by：Huang Yusong

果 Fruits
摄影：黄俞淞 Photo by：Huang Yusong

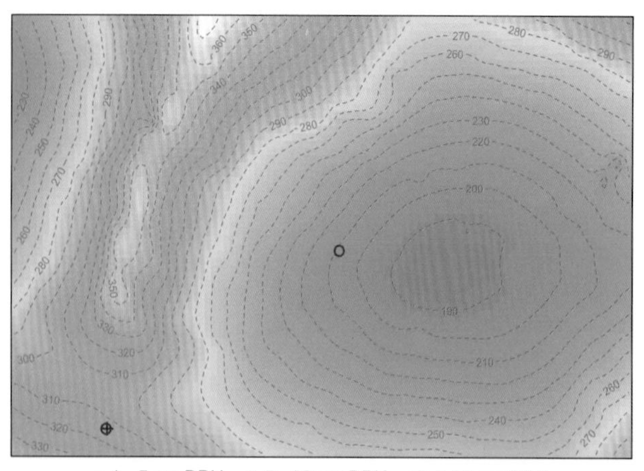

• 1~5 cm DBH + 5~20 cm DBH ○ ≥20 cm DBH
个体分布图 Distribution of individuals

径级分布表 DBH class

胸径区间 (Diameter class) (cm)	个体数 (No. of individuals in the plot)	比例 (Proportion) (%)
1~2	0	0.00
2~5	0	0.00
5~10	1	33.33
10~20	0	0.00
20~35	0	0.00
35~60	1	33.33
≥60	1	33.33

44 卵叶铁苋菜

luǎn yè tiě xiàn cài | Kerr's Copperleaf

Acalypha kerrii Craib
大戟科 Euphorbiaceae

代码（SpCode）= ACAKER
个体数（Individual number/15 hm^2）= 686
最大胸径（Max DBH）= 6.9 cm
重要值排序（Importance value rank）= 29

落叶灌木，高 1~3 m。嫩枝被黄色微柔毛。叶卵形或长卵形，先端渐尖，基出脉 3 条。花单性同序或异序，花序梗很短。蒴果具 3 个分果，果皮具疏毛和数枚短刺。花期 5~6 月。

Deciduous shrubs, 1–3 m tall. Branchlets yellowish pubescent when young. Leaf blade ovate or long ovate, apex acuminate, basal veins 3. Flowers unisexual, androgynous or not, peduncle very short. Capsule 3-locular, pilose and shortly softly few echinate. Fl. May–Jun..

枝叶 Branch and leaves
摄影：黄俞淞 Photo by：Huang Yusong

果序 Infructescence
摄影：黄俞淞 Photo by：Huang Yusong

果枝 Fruiting branch
摄影：黄俞淞 Photo by：Huang Yusong

• 1~5 cm DBH + 5~20 cm DBH ○ ≥20 cm DBH

个体分布图 Distribution of individuals

径级分布表 DBH class

胸径区间 (Diameter class) (cm)	个体数 (No. of individuals in the plot)	比例 (Proportion) (%)
1~2	433	63.12
2~5	244	35.57
5~10	9	1.31
10~20	0	0.00
20~35	0	0.00
35~60	0	0.00
≥60	0	0.00

45 红背山麻杆

hóng bèi shān má gǎn | Red-back Alchornea

Alchornea trewioides (Benth.) Muell. Arg.
大戟科 Euphorbiaceae

代码（SpCode）= ALCTRE
个体数（Individual number/15 hm^2）= 10
最大胸径（Max DBH）= 2.5 cm
重要值排序（Importance value rank）= 164

常绿灌木，高 1~3 m。叶阔卵圆形，背面常呈暗红色，有柔毛，基出脉 3 条，基部有 5 枚红色腺体和 2 枚线性附属体。雌花序顶生，雄花序腋生且为总状花序。蒴果近球形，被灰色柔毛。花期 3~5 月，果期 6~8 月。

Evergreen shrubs, 1–3 m tall. Leaf blade broadly ovate, abaxially dark red, puberulent, basal veins 3, with 5 red glands and 2 filamentose appendage. Female inflorescences terminal, male inflorescences racemose, axillary. Capsule subglobose, gray puberulent. Fl. Mar.–May, fr. Jun.–Aug..

植株 Whole plant
摄影：黄俞淞 Photo by：Huang Yusong

果序 Infructescence
摄影：黄俞淞 Photo by：Huang Yusong

果枝 Fruiting branch
摄影：黄俞淞 Photo by：Huang Yusong

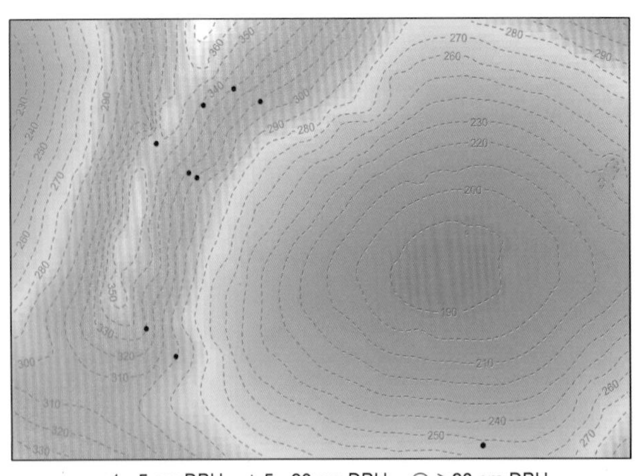

• 1~5 cm DBH + 5~20 cm DBH ○ ≥20 cm DBH
个体分布图 Distribution of individuals

径级分布表 DBH class

胸径区间 (Diameter class) (cm)	个体数 (No. of individuals in the plot)	比例 (Proportion) (%)
1~2	7	70.00
2~5	3	30.00
5~10	0	0.00
10~20	0	0.00
20~35	0	0.00
35~60	0	0.00
≥60	0	0.00

46 五月茶 wǔ yuè chá | Bignay

Antidesma bunius (Linn.) Spreng.
大戟科 Euphorbiaceae

代码（SpCode）= ANTBUN
个体数（Individual number/15 hm^2）= 4
最大胸径（Max DBH）= 3.5 cm
重要值排序（Importance value rank）= 182

常绿灌木或乔木，高可达 10 m。小枝无毛。叶纸质，顶端圆形或急尖，具短尖头，叶柄长 3~10 mm，略被短柔毛。雄花序为顶生或侧生的穗状花序，雌花序为总状花序，顶生。核果近球形，成熟时红色。花期 3~5 月，果期 6~11 月。

Evergreen shrubs or trees, up to 10 m tall. Branchlets glabrous. Leaf blade papery, apex acute to rounded, usually mucronate, petiole 3–10 mm long, slightly short pubescence. Male inflorescences spicate, terminal or lateral, female inflorescences racemose, terminal. Drupes subglobose, red when maturity. Fl. Mar.–May, fr. Jun.–Nov..

果枝 Fruiting branches
摄影：黄俞淞 Photo by：Huang Yusong

果序 Infructescence
摄影：黄俞淞 Photo by：Huang Yusong

叶 Leaves
摄影：黄俞淞 Photo by：Huang Yusong

• 1~5 cm DBH + 5~20 cm DBH ○ ≥20 cm DBH
个体分布图 Distribution of individuals

径级分布表 DBH class

胸径区间 (Diameter class) (cm)	个体数 (No. of individuals in the plot)	比例 (Proportion) (%)
1~2	1	25.00
2~5	3	75.00
5~10	0	0.00
10~20	0	0.00
20~35	0	0.00
35~60	0	0.00
≥60	0	0.00

47 日本五月茶

rì běn wǔ yuè chá | Japan Bignay

Antidesma japonicum Sieb. et Zucc.
大戟科 Euphorbiaceae

代码（SpCode）= ANTJAP
个体数（Individual number/15 hm^2）= 2535
最大胸径（Max DBH）= 38.0 cm
重要值排序（Importance value rank）= 9

常绿灌木或小乔木，高 2~8 m。小枝和叶脉初时被短柔毛。叶纸质，顶端具尾状尖，或渐尖有小尖头。雄花序为分枝或不分枝的总状花序。核果卵圆形。花期 4~7 月，果期 9~11 月。

Evergreen shrubs or small trees, 2–8 m tall. Branchlet and veins pubescent when young. Leaves papyraceous, apex acuminate, caudate, mucronulate. Male flower racemiform, branched or ramous. Drupes ellipsoid. Fl. Apr.–Jul., fr. Sep.–Nov..

树干 Trunk
摄影：王斌 Photo by：Wang Bin

果序 Infructescence
摄影：黄俞淞 Photo by：Huang Yusong

果枝 Fruiting branch
摄影：黄俞淞 Photo by：Huang Yusong

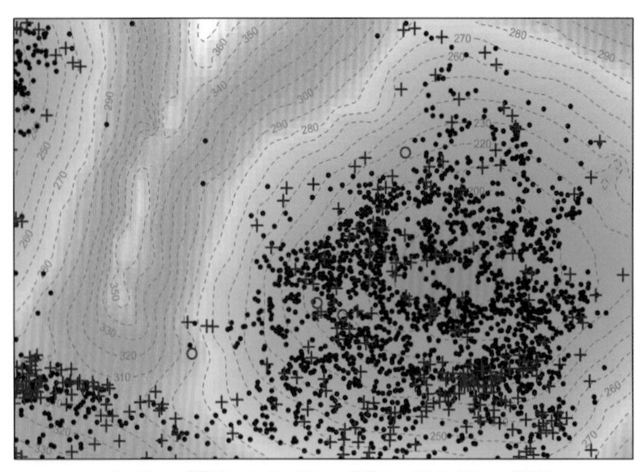

• 1~5 cm DBH + 5~20 cm DBH ○ ⩾20 cm DBH
个体分布图 Distribution of individuals

径级分布表 DBH class

胸径区间 (Diameter class) (cm)	个体数 (No. of individuals in the plot)	比例 (Proportion) (%)
1~2	1063	41.93
2~5	1157	45.64
5~10	256	10.10
10~20	55	2.17
20~35	3	0.12
35~60	1	0.04
⩾60	0	0.00

48 山地五月茶

shān dì wǔ yuè chá | Mountain Bignay

Antidesma montanum Bl.
大戟科 Euphorbiaceae

代码（SpCode）= ANTMON
个体数（Individual number/15 hm^2）= 61
最大胸径（Max DBH）= 9.3 cm
重要值排序（Importance value rank）= 109

灌木或小乔木。叶片纸质，长 7~25 cm，宽 2~10 cm，先端具长或短的尾状尖，或渐尖有小尖头，托叶线形，长 4~10 mm。总状花序顶生或腋生，分枝或不分枝。核果卵圆形。花期 4~7 月，果期 7~11 月。

Shrubs or small trees. Leaves papyraceous, 7–25 cm long, 2–10 cm wide, apex with long or short caudate cusp, or acumen, stipule linear, 4–10 mm long. Inflorescences axillary or terminal, branched or unbranched. Drupes ellipsoid. Fl. Apr.–Jul., fr. Jul.–Nov..

树干 Trunk
摄影：王斌 Photo by：Wang Bin

叶 Leaves
摄影：吴望辉 Photo by：Wu Wanghui

果序 Infructescence
摄影：吴望辉 Photo by：Wu Wanghui

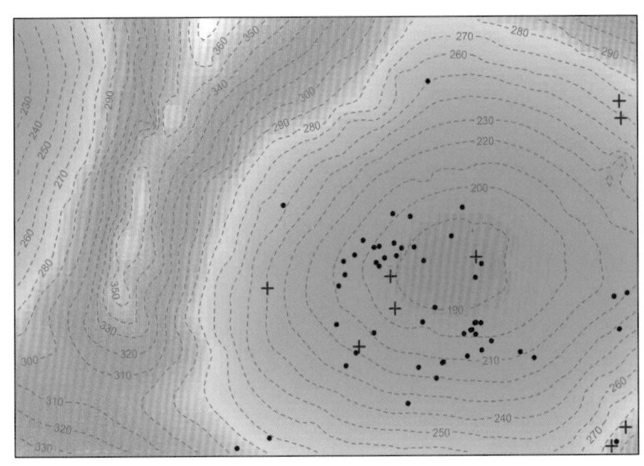

• 1~5 cm DBH + 5~20 cm DBH ○ ≥20 cm DBH
个体分布图 Distribution of individuals

径级分布表 DBH class

胸径区间 (Diameter class) (cm)	个体数 (No. of individuals in the plot)	比例 (Proportion) (%)
1~2	29	47.54
2~5	23	37.70
5~10	9	14.75
10~20	0	0.00
20~35	0	0.00
35~60	0	0.00
≥60	0	0.00

49 木奶果

mù nǎi guǒ | Ramiflorous Baccaurea

Baccaurea ramiflora Lour.
大戟科 Euphorbiaceae

代码（SpCode）= BACRAM
个体数（Individual number/15 hm^2）= 41
最大胸径（Max DBH）= 9.2 cm
重要值排序（Importance value rank）= 123

常绿乔木，高 5~15 m。树皮灰褐色。叶片纸质，先端短渐尖至急尖，两面无毛，侧脉每边 5~7 条。花小，雌雄异株，无花瓣，总状圆锥花序腋生或茎生，被疏短柔毛。蒴果浆果状，卵形或近球形，黄色，后变紫红色。花期 3~4 月，果期 6~10 月。

Evergreen trees, 5–15 m tall. Bark gray-brown. Leaf blade papyraceous, apex shortly acuminate to acute, glabrous on both surfaces, lateral veins 5–7 paris. Flowers small, dioecious, apetalous, racemelike panicles axillary or cauline, thinly puberulent. Capsules baccate, ovaid or subglobose, yellow, red when mature. Fl. Mar.–Apr., fr. Jun.–Oct..

果序 Infructescence
摄影：黄俞淞 Photo by：Huang Yusong

果枝 Fruiting branch
摄影：黄俞淞 Photo by：Huang Yusong

叶 Leaves
摄影：黄俞淞 Photo by：Huang Yusong

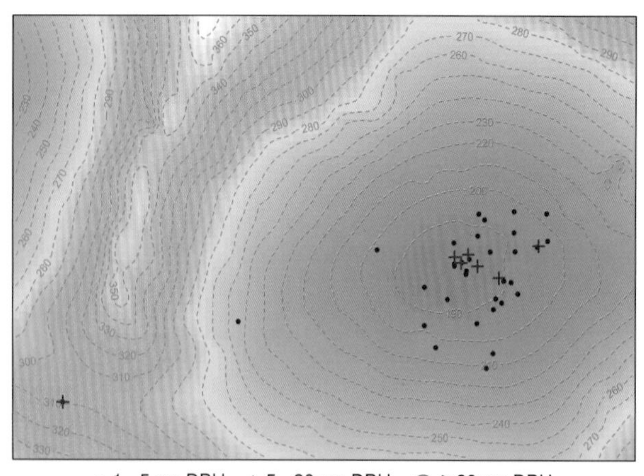

• 1~5 cm DBH + 5~20 cm DBH ○ ≥20 cm DBH
个体分布图 Distribution of individuals

径级分布表 DBH class

胸径区间 (Diameter class) (cm)	个体数 (No. of individuals in the plot)	比例 (Proportion) (%)
1~2	7	17.07
2~5	27	65.85
5~10	7	17.07
10~20	0	0.00
20~35	0	0.00
35~60	0	0.00
≥60	0	0.00

50 秋枫

qiū fēng | Javanese bishopweed

Bischofia javanica Blume
大戟科 Euphorbiaceae

代码（SpCode）= BISJAV
个体数（Individual number/15 hm^2）= 197
最大胸径（Max DBH）= 35.9 cm
重要值排序（Importance value rank）= 53

常绿或半常绿乔木，高可达 40 m。树皮灰褐色至棕褐色，厚约 1 cm，具红色汁液。三出复叶，稀 5 小叶。花小，雌雄异株，多朵组成腋生的圆锥花序。浆果球形，直径 8~15 mm，成熟时褐色。花期 3~4 月，果期 11~12 月。

Evergreen or semievergreen trees, up to 40 m tall. Bark gray-brown to brown, ca. 1 cm thick, with red latex. Leaves palmately 3(5)-foliolate. Flowers small, dioecious, Inflorescence axillary, paniculate. Bacca globose, 8-15 mm in diam., brownish. Fl. Mar.-Apr., fr. Nov.-Dec..

树干 Trunk
摄影：王斌 Photo by: Wang Bin

果枝 Fruiting branch
摄影：刘晟源 Photo by: Liu Shengyuan

果 Fruits
摄影：黄俞淞 Photo by: Huang Yusong

• 1~5 cm DBH + 5~20 cm DBH ○ ≥20 cm DBH

个体分布图 Distribution of individuals

径级分布表 DBH class

胸径区间 (Diameter class) (cm)	个体数 (No. of individuals in the plot)	比例 (Proportion) (%)
1~2	11	5.58
2~5	31	15.74
5~10	82	41.62
10~20	59	29.95
20~35	12	6.09
35~60	2	1.02
≥60	0	0.00

51 黑面神

hēi miàn shén | Fruticosa Breynia

Breynia fruticosa (Linn.) Hook. f.
大戟科 Euphorbiaceae

代码（SpCode）= BREFRU
个体数（Individual number/15 hm^2）= 2
最大胸径（Max DBH）= 1.2 cm
重要值排序（Importance value rank）= 200

灌木，高 1~3 m。茎灰褐色。全株均无毛。叶片革质，卵形、阔卵形，叶面深绿色，叶背粉绿色，干后变黑色，具有小斑点。花小，单生或 2~4 朵簇生于叶腋内，雌花位于小枝上部，雄花位于小枝下部。蒴果圆球状，有宿存花萼。花期 4~9 月，果期 5~12 月。

Shrubs, 1–3 m tall. Stem gray-brown. Glabrous throughout. Leaf blade leathery, apex obtuse or acute, adaxially dark green, abaxially pink green, black when dry, with small macula. Flower samll, solitary or 2–4-flowered in axillary clusters, male in proximal axils, female in distal axils. Capsules globose, with persistent calyx. Fl. Apr.–Sep, fr. May–Dec..

枝叶 Branches and leaves
摄影：黄俞凇 Photo by：Huang Yusong

果枝 Fruiting branch
摄影：黄俞凇 Photo by：Huang Yusong

雌花枝 Female flowering branch
摄影：丁涛 Photo by：Ding Tao

• 1~5 cm DBH + 5~20 cm DBH ○ ≥20 cm DBH
个体分布图 Distribution of individuals

径级分布表 DBH class

胸径区间 (Diameter class) (cm)	个体数 (No. of individuals in the plot)	比例 (Proportion) (%)
1~2	2	100.00
2~5	0	0.00
5~10	0	0.00
10~20	0	0.00
20~35	0	0.00
35~60	0	0.00
≥60	0	0.00

52 禾串树

hé chuàn shù | Balansa's Bridelia

Bridelia balansae Tutcher
大戟科 Euphorbiaceae

代码（SpCode）= BRIBAL
个体数（Individual number/15 hm^2）= 125
最大胸径（Max DBH）= 18.5 cm
重要值排序（Importance value rank）= 74

小乔木或乔木。树皮黄褐色，小枝具有凸起的皮孔，无毛。叶片近革质，椭圆形或长椭圆形，边缘反卷，托叶线状披针形，长约 3 mm。花雌雄同序，密集成腋生的团伞花序。核果长卵形，成熟时紫黑色。花期 3~8 月，果期 9~11 月。

Small trees or trees. Bark yellow-brown, branchlet has salient lenticel, glabrous. Leaf blade nearly leathery, elliptic or elliptic-lanceolate, margin slightly revolute, stipule linear-lanceolate, ca. 3 mm long. Flowers monoecious, glomerules axillary. Drupe oblong, atropurpurea. Fl. Mar.–Aug., fr. Sep.–Nov..

树干 Trunk
摄影：黄俞淞 Photo by：Huang Yusong

果序 Infructescence
摄影：黄俞淞 Photo by：Huang Yusong

枝叶 Branch and leaves
摄影：黄俞淞 Photo by：Huang Yusong

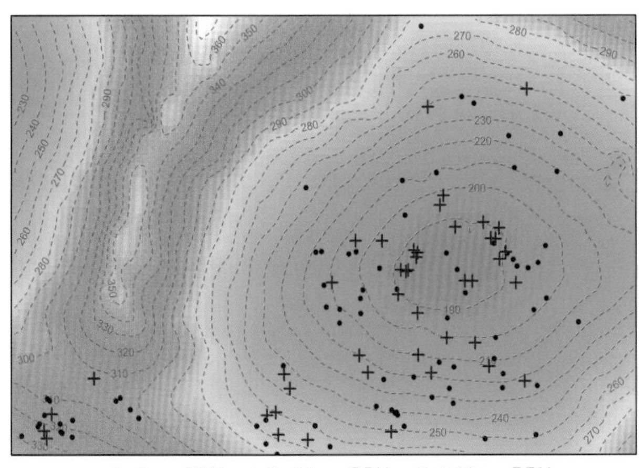

• 1~5 cm DBH + 5~20 cm DBH ○ ≥20 cm DBH
个体分布图 Distribution of individuals

径级分布表 DBH class

胸径区间 (Diameter class) (cm)	个体数 (No. of individuals in the plot)	比例 (Proportion) (%)
1~2	41	32.80
2~5	40	32.00
5~10	34	27.20
10~20	10	8.00
20~35	0	0.00
35~60	0	0.00
≥60	0	0.00

53 膜叶土蜜树

mó yè tǔ mì shù | Glaucous Bridelia

Bridelia glauca Blume
大戟科 Euphorbiaceae

代码（SpCode）= BRIGLA
个体数（Individual number/15 hm^2）= 20
最大胸径（Max DBH）= 16.2 cm
重要值排序（Importance value rank）= 147

乔木，高达 15 m。叶片膜质或近膜质，倒卵形、长圆形或椭圆状卵形，先端急尖至渐尖，基部钝至圆，侧脉每边 7~12 条。花白色，雌雄同株，多朵簇生于叶腋内或组成穗状花序。核果椭圆状，顶端具小尖头，基部有宿存萼片。花期 5~9，果期 9~12 月。

Trees, up to 15 m tall. Leaf blade membranous or thickly papery, elliptic-ovate, oblong, or obovate, apex acute to acuminate, base obtuse to round, lateral veins 7–12 paris. Flowers white, monoecious, many flowers fascicles axillary, or spicate. Drupes ellipsoidal, apex with mucro, base with persistent sepals. Fl. May–Sep., fr. Sep.–Dec..

果枝 Fruiting branches
摄影：谭运洪 Photo by：Tan Yunhong

果序 Infructescence
摄影：谭运洪 Photo by：Tan Yunhong

叶背 Leaf back
摄影：谭运洪 Photo by：Tan Yunhong

• 1~5 cm DBH + 5~20 cm DBH ○ ≥20 cm DBH
个体分布图 Distribution of individuals

径级分布表 DBH class

胸径区间 (Diameter class) (cm)	个体数 (No. of individuals in the plot)	比例 (Proportion) (%)
1~2	6	30.00
2~5	6	30.00
5~10	5	25.00
10~20	3	15.00
20~35	0	0.00
35~60	0	0.00
≥60	0	0.00

54 土密树

tǔ mì shù | Tomentose Bridelia

Bridelia tomentosa Blume
大戟科 Euphorbiaceae

代码（SpCode）= BRITOM
个体数（Individual number/15 hm^2）= 15
最大胸径（Max DBH）= 8.5 cm
重要值排序（Importance value rank）= 146

常绿乔木，高可达 17 m。树皮灰褐色，近平滑。叶片近革质，边缘反卷，托叶线状披针形，长约 3 mm，被黄色柔毛。花雌雄同序，密集成腋生的团伞花序。核果长卵形，成熟时紫黑色。花期 3~8 月，果期 9~11 月。

Evergreen trees, up to 17 m tall. Bark gray-brown, nearly glabrous. Leaf blade nearly leathery, margin revolute, stipule linear-lanceolate, ca. 3 mm long, yellow pubescence. Flowers monoecious, densely fascicles axillary. Drupes long ovate, atropurpureus when mature. Fl. Mar.–Aug., fr. Sep.–Nov..

嫩枝 New branch
摄影：黄俞淞 Photo by：Huang Yusong

枝叶 Branch and leaves
摄影：黄俞淞 Photo by：Huang Yusong

叶 Leaf
摄影：黄俞淞 Photo by：Huang Yusong

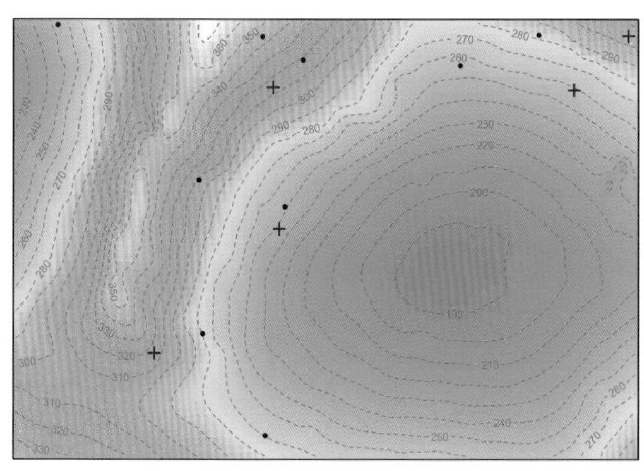

• 1~5 cm DBH + 5~20 cm DBH ○ ⩾20 cm DBH
个体分布图 Distribution of individuals

径级分布表 DBH class

胸径区间 (Diameter class) (cm)	个体数 (No. of individuals in the plot)	比例 (Proportion) (%)
1~2	6	40.00
2~5	4	26.67
5~10	5	33.33
10~20	0	0.00
20~35	0	0.00
35~60	0	0.00
⩾60	0	0.00

55 肥牛树

féi niú shù | Chinese Cephalomappa

Cephalomappa sinensis (Chun et F. C. How) Kosterm.
大戟科 Euphorbiaceae

代码（SpCode）= CEPSIN
个体数（Individual number/15 hm^2）= 917
最大胸径（Max DBH）= 42.4 cm
重要值排序（Importance value rank）= 17

常绿乔木，高 20 m。叶长椭圆形或长倒卵形，基部歪斜，具两枚小腺体，叶缘浅波状或有疏齿，网脉明显。花单性同株，雄花位于花序顶部，雌花 1 朵至数朵生于花序基部。蒴果具密集的小疣状突起。花期 3~4 月，果期 6~7 月。

Evergreen trees, up to 20 m tall. Leaf blade oblong or long obovate, base skew, with two small gland, margin entire to crenate or dentate, net-vein obvious. Flowers imperfect, bisexual, male flowers in mostly terminal clusters, female flowers 1 to few at base. Capsule densely verrucose. Fl. Mar.–Apr., fr. Jun.–Jul..

树干 Trunk
摄影：王斌 Photo by: Wang Bin

枝叶 Branch and leaves
摄影：王斌 Photo by: Wang Bin

果 Fruits
摄影：黄俞淞 Photo by: Huang Yusong

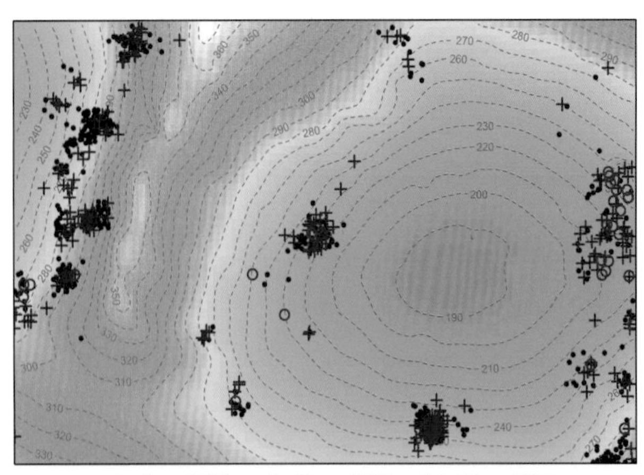

• 1~5 cm DBH + 5~20 cm DBH ○ ≥20 cm DBH
个体分布图 Distribution of individuals

径级分布表 DBH class

胸径区间 (Diameter class) (cm)	个体数 (No. of individuals in the plot)	比例 (Proportion) (%)
1~2	297	32.39
2~5	315	34.35
5~10	171	18.65
10~20	104	11.34
20~35	28	3.05
35~60	2	0.22
≥60	0	0.00

56 白桐树 bái tóng shù | Indian Claoxylon

Claoxylon indicum (Reinw. ex Bl.) Hassk
大戟科 Euphorbiaceae

代码（SpCode）= CLAIND
个体数（Individual number/15 hm^2）= 1
最大胸径（Max DBH）= 29.6 cm
重要值排序（Importance value rank）= 183

小乔木或灌木，高 3~12 m。叶片纸质，通常卵形或卵圆形，两面均被疏毛，边缘具不规则的小齿或锯齿；叶柄长 5~15 cm，顶部具 2 枚小腺体。雌雄异株，花序各部均被绒毛。蒴果具 3 个分果爿，被灰色短茸毛。花果期 3~12 月。

Small trees or shrubs, 3–12 m tall. Leaf blade papery, usually ovate to broadly ovate, both surfaces pilose, margin crenulate or dentate, petilole 5–15 cm long, with 2 glands. Dioecious, inflorescences tomentulose. Capsule 3-lobed. grayly tomentulose. Fl. and fr. Mar.–Dec..

果枝 Fruiting branch
摄影：黄俞淞 Photo by：Huang Yusong

枝叶 Branch and leaves
摄影：黄俞淞 Photo by：Huang Yusong

果序 Infructescence
摄影：黄俞淞 Photo by：Huang Yusong

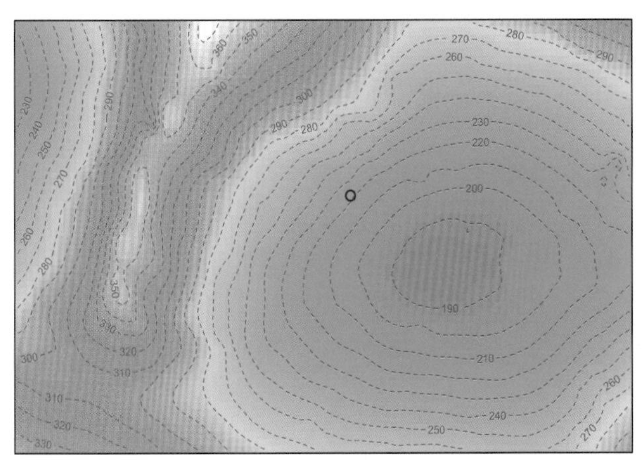

• 1~5 cm DBH + 5~20 cm DBH ○ ≥20 cm DBH
个体分布图 Distribution of individuals

径级分布表 DBH class

胸径区间 (Diameter class) (cm)	个体数 (No. of individuals in the plot)	比例 (Proportion) (%)
1~2	0	0.00
2~5	0	0.00
5~10	0	0.00
10~20	0	0.00
20~35	1	100.00
35~60	0	0.00
≥60	0	0.00

57 假肥牛树

jiǎ féi niú shù | Petelot's Cleistanthus

Cleistanthus petelotii Merr. ex Croizat
大戟科 Euphorbiaceae

代码（SpCode）= CLEPET
个体数（Individual number/15 hm^2）= 613
最大胸径（Max DBH）= 26.8 cm
重要值排序（Importance value rank）= 52

常绿乔木，高 7~18 m。叶革质，边缘略反卷，侧脉未达叶缘而结。花雌雄同株，雄花无毛，雌花数朵组成腋生团伞花序。蒴果近圆球形，外果皮具网状皱纹。花期 3~4 月，果期 7~9 月。

Evergreen trees, 7–18 m tall. Leaf blade leathery, margin slightly volume, lateral veins anastomosing near margin. Flowers monoecious, male glabrous, female several grouped into axillary glomerules. Capsules subglobose, epicarps reticulate-rugose. Fl. Mar.–Apr., fr. Jul.–Sep..

果枝 Fruiting branch
摄影：黄俞淞 Photo by: Huang Yusong

叶 Leaves
摄影：黄俞淞 Photo by: Huang Yusong

叶背 Leaf back
摄影：丁涛 Photo by: Ding Tao

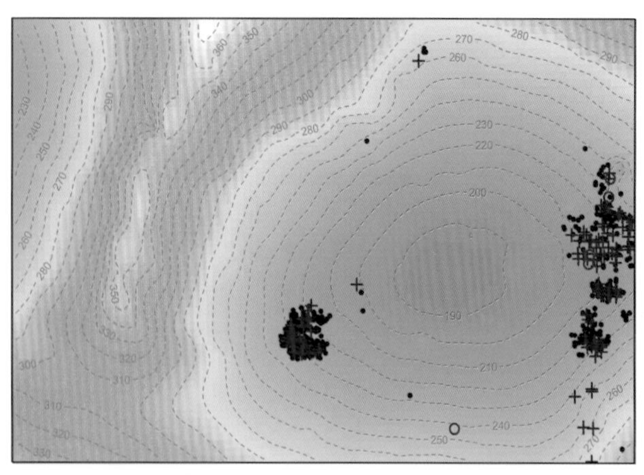

• 1~5 cm DBH + 5~20 cm DBH ○ ≥20 cm DBH
个体分布图 Distribution of individuals

径级分布表 DBH class

胸径区间 (Diameter class) (cm)	个体数 (No. of individuals in the plot)	比例 (Proportion) (%)
1~2	252	41.11
2~5	232	37.85
5~10	84	13.70
10~20	40	6.53
20~35	5	0.82
35~60	0	0.00
≥60	0	0.00

58 闭花木

bì huā mù | Sumatran Cleistanthus

Cleistanthus sumatranus (Miq.) Muell. Arg.
大戟科 Euphorbiaceae

代码（SpCode）= CLESUM
个体数（Individual number/15 hm^2）= 9977
最大胸径（Max DBH）= 55.0 cm
重要值排序（Importance value rank）= 1

常绿灌木或乔木，高 3~10 m。树皮红褐色。叶纸质，先端尾状渐尖。花雌雄同株，单生或簇生于叶腋。蒴果卵状三棱形，果皮薄而脆，成熟时分裂成 3 分果爿。花期 3~8 月，果期 4~10 月。

Evergreen shurbs or trees, 3–10 m tall. Bark red-brown. Leaf blade papery, apex caudate acuminate. Flowers monoecious, axillary, one or few-flowered fascicles. Capsules ovoid-trigonous, carpodermis thin, 3-valved when mature. Fl. Mar.–Aug, fr. Apr.–Oct..

树干 Trunk
摄影：王斌 Photo by：Wang Bin

枝叶 Branch and leaves
摄影：黄俞淞 Photo by：Huang Yusong

果 Fruits
摄影：黄俞淞 Photo by：Huang Yusong

• 1~5 cm DBH + 5~20 cm DBH ○ ≥20 cm DBH
个体分布图 Distribution of individuals

径级分布表 DBH class

胸径区间 (Diameter class) (cm)	个体数 (No. of individuals in the plot)	比例 (Proportion) (%)
1~2	3063	30.70
2~5	3586	35.94
5~10	2072	20.77
10~20	1155	11.58
20~35	99	0.99
35~60	2	0.02
≥60	0	0.00

59 石山巴豆

shí shān bā dòu | Broad-leaf Croton

Croton euryphyllus W. W. Sm.
大戟科 Euphorbiaceae

代码（SpCode）= CROEUR
个体数（Individual number/15 hm^2）= 422
最大胸径（Max DBH）= 23.1 cm
重要值排序（Importance value rank）= 43

落叶灌木或小乔木，高 3~5 m。嫩枝、叶和花序均被星状柔毛。叶纸质，近圆形至阔卵形，叶柄顶端有 2 枚具柄腺体。总状花序。蒴果近圆球状，密被短星状毛。花期 4~5 月。

Deciduous shurbs or small trees, 3–5 m tall. Twig, leaf and inflorescence stellate-pubescent. Leaf blade papery, nearly round to broadly ovate, apex of petiole with 2 stalked glands. Racemose. Capsules subglobose, densely stellate-pubescent. Fl. Apr.–May.

树干 Trunk
摄影：王斌 Photo by: Wang Bin

果枝 Fruiting branches
摄影：黄俞淞 Photo by: Huang Yusong

花枝 Flowering branches
摄影：黄俞淞 Photo by: Huang Yusong

• 1~5 cm DBH + 5~20 cm DBH ○ ≥20 cm DBH
个体分布图 Distribution of individuals

径级分布表 DBH class

胸径区间 (Diameter class) (cm)	个体数 (No. of individuals in the plot)	比例 (Proportion) (%)
1~2	146	34.60
2~5	177	41.94
5~10	63	14.93
10~20	33	7.82
20~35	3	0.71
35~60	0	0.00
≥60	0	0.00

60 密花核果木 mì huā hé guǒ mù | Dense-flower Drypetes

Drypetes congestiflora Chun et T. C. Chen
大戟科 Euphorbiaceae

代码（SpCode）= DRYCON
个体数（Individual number/15 hm^2）= 710
最大胸径（Max DBH）= 51.1 cm
重要值排序（Importance value rank）= 23

常绿乔木，高 6~22 m。树皮灰褐色。叶革质，基部多少偏斜，边缘明显锯齿状。雄花密集成簇。蒴果圆球形。花期 2~7，果期 6~10 月。

Evergreen trees, 6–22 m tall. Bark gray-brown. Leaf blade leathery, base more or less deflective, margin prominently serrate. Male flowers densely clustered. Capsule globose. Fl. Feb.–Jul., fr. Jun.–Oct..

树干 Trunk
摄影：王斌 Photo by：Wang Bin

叶 Leaves
摄影：王斌 Photo by：Wang Bin

叶背 Leaf back
摄影：王斌 Photo by：Wang Bin

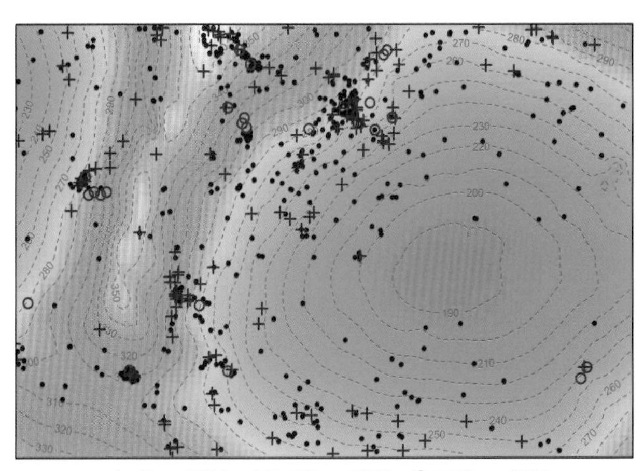

• 1~5 cm DBH + 5~20 cm DBH ○ ≥20 cm DBH
个体分布图 Distribution of individuals

径级分布表 DBH class

胸径区间 (Diameter class) (cm)	个体数 (No. of individuals in the plot)	比例 (Proportion) (%)
1~2	268	37.75
2~5	273	38.45
5~10	99	13.94
10~20	47	6.62
20~35	21	2.96
35~60	2	0.28
≥60	0	0.00

61 网脉核果木

wǎng mài hé guǒ mù | Very-netveined Drypetes

Drypetes perreticulata Gagnep.
大戟科 Euphorbiaceae

代码（SpCode）= DRYPER
个体数（Individual number/15 hm^2）= 631
最大胸径（Max DBH）= 48.7 cm
重要值排序（Importance value rank）= 21

常绿乔木，高 12~16 m。树皮灰黄色，平滑，小枝具棱。叶革质，叶基部两侧不相等，网脉在两面明显凸起。雄花 2~3 朵簇生于叶腋，雌花腋生，柱头 2 裂。核果单生于叶腋，成熟时红色。花期 1~3 月，果期 5~10 月。

Evergreen trees, 12–16 m tall. Bark gray-yellow, glabrous, twig angulate. Leaf leathery, base oblique, net-veins obviously prominent. Male flowers axillary, usually 2 or 3 together, female flowers axillary, stigma 2-lobed. Drupes axillary, red when mature. Fl. Jan.–Mar., fr. May–Oct..

树干 Trunk
摄影：王斌 Photo by：Wang Bin

果枝 Fruiting branches
摄影：黄俞淞 Photo by：Huang Yusong

叶背 Leaf back
摄影：王斌 Photo by：Wang Bin

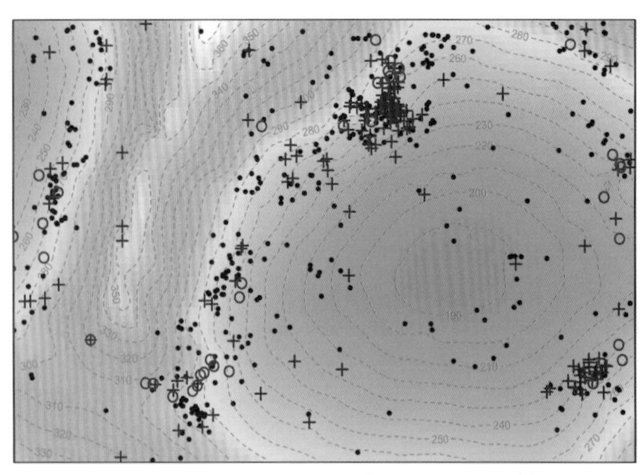

• 1~5 cm DBH + 5~20 cm DBH ○ ≥20 cm DBH
个体分布图 Distribution of individuals

径级分布表 DBH class

胸径区间 (Diameter class) (cm)	个体数 (No. of individuals in the plot)	比例 (Proportion) (%)
1~2	234	37.08
2~5	193	30.59
5~10	79	12.52
10~20	87	13.79
20~35	34	5.39
35~60	4	0.63
≥60	0	0.00

62 白饭树

bái fàn shù | Stinking Flueggea

Flueggea virosa (Roxburgh ex Willdenow) Voigt
大戟科 Euphorbiaceae

代码（SpCode）= FLUVIR
个体数（Individual number/15 hm^2）= 46
最大胸径（Max DBH）= 13.0 cm
重要值排序（Importance value rank）= 116

灌木，高 1~6 m。叶片纸质，先端圆至急尖，全缘，下面白绿色。花小，淡黄色，雌雄异株，多朵簇生于叶腋。蒴果浆果状，近于圆形，成熟时果皮淡白色。花期 3~8 月，果期 7~12 月。

Shurbs, 1–6 m tall. Leaf blade papery, apex rounded to acute, entire, abaxial white-green. Flowers small. dioecious, axillary, fascicled. Berry nearly rounded, whitish when mature. Fl. Mar.–Aug., fr. Jul.–Dec..

果枝 Fruiting branches
摄影：黄俞淞 Photo by：Huang Yusong

叶 Leaves
摄影：黄俞淞 Photo by：Huang Yusong

果 Fruits
摄影：刘晟源 Photo by：Liu Shengyuan

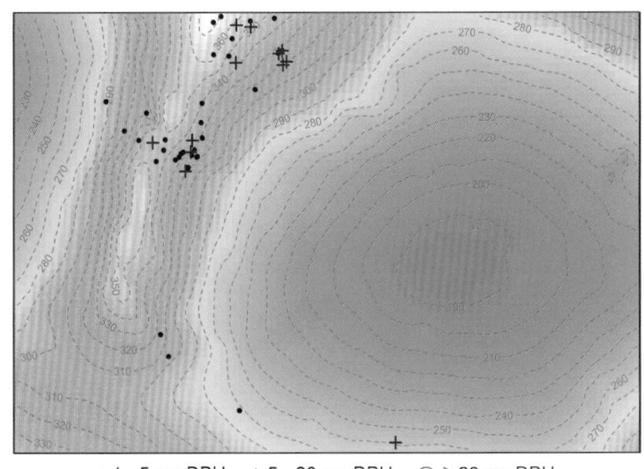

• 1~5 cm DBH + 5~20 cm DBH ○ ≥20 cm DBH
个体分布图 Distribution of individuals

径级分布表 DBH class

胸径区间 (Diameter class) (cm)	个体数 (No. of individuals in the plot)	比例 (Proportion) (%)
1~2	11	23.91
2~5	23	50.00
5~10	10	21.74
10~20	2	4.35
20~35	0	0.00
35~60	0	0.00
≥60	0	0.00

63 红算盘子

hóng suàn pán zǐ | Scarlet Glochidion

Glochidion coccineum (Buch.-Ham.) Muell.
大戟科 Euphorbiaceae

代码（SpCode）= GLOCOC
个体数（Individual number/15 hm^2）= 27
最大胸径（Max DBH）= 14.7 cm
重要值排序（Importance value rank）= 127

常绿灌木或小乔木，通常高约 4 m。枝条具棱，被短柔毛。叶片革质，长 6~12 cm，宽 3~5 cm，叶柄和托叶被柔毛。花 2~6 朵簇生于叶腋内。蒴果扁球状，直径约 15 mm，有 10 条纵沟，被微毛。花期 4~10 月，果期 8~12 月。

Evergreen shurbs or small trees, usually ca. 4 m tall. Branches angular, pubescent. Leaf blade leathery, 6–12 cm long, 3–5 cm wide, petiole and stipule pubescent. Flowers in axillary clusters. Capsules depressed globose, about 15 mm in diam., 10-grooved, puberulent. Fl. Apr.–Oct., fr. Aug.–Dec..

树干 Trunk
摄影：王斌 Photo by：Wang Bin

枝叶 Branch and leaves
摄影：黄俞淞 Photo by：Huang Yusong

叶 Leaves
摄影：王斌 Photo by：Wang Bin

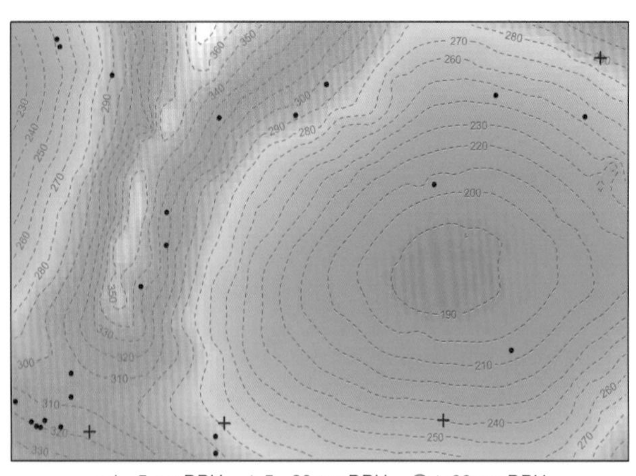

• 1~5 cm DBH + 5~20 cm DBH ○ ≥20 cm DBH
个体分布图 Distribution of individuals

径级分布表 DBH class

胸径区间 (Diameter class) (cm)	个体数 (No. of individuals in the plot)	比例 (Proportion) (%)
1~2	10	37.04
2~5	13	48.15
5~10	3	11.11
10~20	1	3.70
20~35	0	0.00
35~60	0	0.00
≥60	0	0.00

64 四裂算盘子 sì liè suàn pán zǐ | Elliptical Glochidion

Glochidion ellipticum Wight
大戟科 Euphorbiaceae

代码（SpCode）= GLOELL
个体数（Individual number/15 hm^2）= 35
最大胸径（Max DBH）= 19.0 cm
重要值排序（Importance value rank）= 122

灌木或小乔木，高可达 10 m。叶片纸质或近革质，长 9~15 cm，宽 3.5~4.5 cm，侧脉每边 6~8 条。多朵雄花与少数几朵雌花同时簇生于叶腋内。蒴果扁球状，直径 6~8 mm，具 3~5 纵棱。花期 3 月 ~8 月，果期 7 月 ~11 月。

Shrubs or small trees, up to 10 m tall. Leaf blade papery or nearly leathery, 9–15 cm long, 3.5–4.5 cm wide, lateral veins 6–8 pairs. Flowers in biesxual axillary clusters, with many male flowers and few female flowers. Capsules depressed globose, 6–8 mm in diam., 3-5-grooved. Fl. Mar.–Aug., fr. Jul.–Nov..

植株 Whole plant
摄影：黄俞淞 Photo by：Huang Yusong

叶 Leaves
摄影：黄俞淞 Photo by：Huang Yusong

枝叶 Branch and leaves
摄影：黄俞淞 Photo by：Huang Yusong

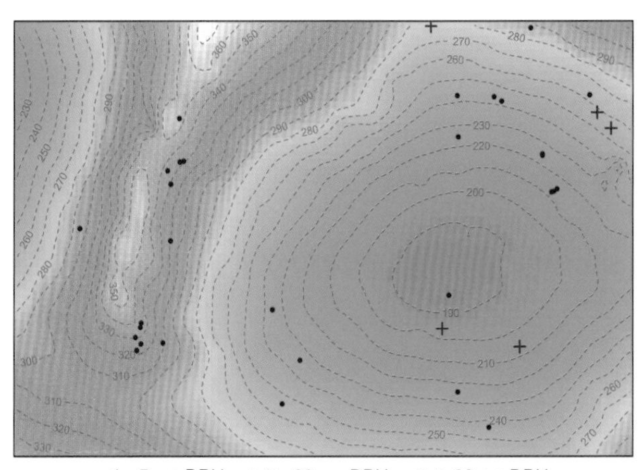

• 1~5 cm DBH + 5~20 cm DBH ○ ⩾20 cm DBH
个体分布图 Distribution of individuals

径级分布表 DBH class

胸径区间 (Diameter class) (cm)	个体数 (No. of individuals in the plot)	比例 (Proportion) (%)
1~2	17	48.57
2~5	13	37.14
5~10	3	8.57
10~20	2	5.71
20~35	0	0.00
35~60	0	0.00
⩾60	0	0.00

65 毛果算盘子

máo guǒ suàn pán zǐ | Woolly-fruit Glochidion

Glochidion eriocarpum Champ. ex Benth.
大戟科 Euphorbiaceae

代码（SpCode）= GLOERI
个体数（Individual number/15 hm^2）= 95
最大胸径（Max DBH）= 12.5 cm
重要值排序（Importance value rank）= 99

灌木，高达 5 m。密被淡黄褐色、扩展的长柔毛。叶片纸质，卵形、狭卵形或宽卵形，两面均被长柔毛，叶柄长 1~2 mm，被柔毛；托叶线形，长 3~4 mm。花单生或 2~4 朵簇生于叶腋内。蒴果扁球状，直径 8~10 mm，具 4~5 条纵沟，密被长柔毛。花果期几乎全年。

Shrubs, up to 5 m tall. Densely xanthic patent villus. Leaf blade papery, ovate, narrowly ovate, or broadly ovate, densely villous on both furface, petiole 1–2 mm, densely villous, stipule linear, 3–4 mm long. Flowers axillary, solitary or 2–4-flowered clusters. Capsules depressed globose, 8–10 mm in diam., 4–5-lobed, densely villous. Fl. and fr. almost seasons.

果枝 Fruiting branch
摄影：黄俞淞 Photo by：Huang Yusong

叶 Leaves
摄影：黄俞淞 Photo by：Huang Yusong

果 Fruits
摄影：黄俞淞 Photo by：Huang Yusong

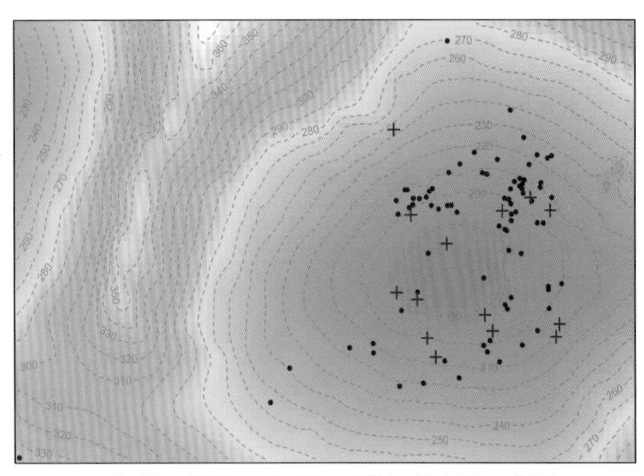

• 1~5 cm DBH + 5~20 cm DBH ○ ≥20 cm DBH
个体分布图 Distribution of individuals

径级分布表 DBH class

胸径区间 (Diameter class) (cm)	个体数 (No. of individuals in the plot)	比例 (Proportion) (%)
1~2	38	40.00
2~5	43	45.26
5~10	13	13.68
10~20	1	1.05
20~35	0	0.00
35~60	0	0.00
≥60	0	0.00

66 艾胶算盘子

ài jiāo suàn pán zǐ | Lanceolate Glochidion

Glochidion lanceolarium (Roxb.) Voigt
大戟科 Euphorbiaceae

代码（SpCode）= GLOLAN
个体数（Individual number/15 hm^2）= 6
最大胸径（Max DBH）= 6.1 cm
重要值排序（Importance value rank）= 173

常绿灌木或小乔木，高 1~3 m。除子房和蒴果外，全株均无毛。叶片革质，长 6~17.5（22）mm，宽 2.5~6.5（8）mm，基部急尖或阔楔形而稍下延。花簇生于叶腋内。蒴果近球状，直径 12~18 mm，边缘具 6~8 条纵沟。花期 4~9 月，果期 7 月至翌年 2 月。

Evergreen shrubs or small trees, 1–3 m tall. Glabrous throughout except for hairy ovary and capsule. Leaf blade leathery, 6–17.5（22）cm long, 2.5–6.5（8）cm wide, base acute or broadly cuneate, slightly decurved. Flowers in axillary clusters. Capsules subglobose, 1.2–1.8 cm in diam., margin 6–8 grooved, Fl. Apr.–Sep., fr. Jul.–Feb.of next year.

枝叶 Branch and leaves
摄影：黄俞淞 Photo by：Huang Yusong

果 Fruits
摄影：黄俞淞 Photo by：Huang Yusong

花序 Inflorescence
摄影：黄俞淞 Photo by：Huang Yusong

• 1~5 cm DBH + 5~20 cm DBH ○ ≥20 cm DBH
个体分布图 Distribution of individuals

径级分布表 DBH class

胸径区间 (Diameter class) (cm)	个体数 (No. of individuals in the plot)	比例 (Proportion) (%)
1~2	3	50.00
2~5	2	33.33
5~10	1	16.67
10~20	0	0.00
20~35	0	0.00
35~60	0	0.00
≥60	0	0.00

67 毛桐

máo tóng | Barbate Mallotus

Mallotus barbatus (Wall.) Muell. Arg.
大戟科 Euphorbiaceae

代码（SpCode）= MALBAR
个体数（Individual number/15 hm^2）= 19
最大胸径（Max DBH）= 7.5 cm
重要值排序（Importance value rank）= 158

小乔木，高 3~8 m。嫩枝、叶和花序均密被黄棕色星状长绒毛。叶纸质，卵状三角形或卵状菱形，掌状脉 5~7 条，近叶柄处具黑色斑状腺体数个。花雌雄异株，总状花序，顶生。蒴果球形，密被淡黄色星状毛和紫红色软刺。花期 4~5 月，果期 9~10 月。

Small trees, 3–8 m tall. Branchlets, leaf, and inflorescences densely brownish stellate-villous. Leaf blade papery, ovate-triangle or ovate-rhomb, palmate vein 5–7, some black gland at the apex of petiole. Inflorescences terminal, diecious, raceme, rarely panicle. Capsule globose, densely stellate-tomentose and softly spiny. Fl. Apr.–May, fr. Sep.–Oct..

花枝 Flowering branches
摄影：黄俞淞 Photo by: Huang Yusong

果序 Infructescence
摄影：黄俞淞 Photo by: Huang Yusong

花序 Inflorescence
摄影：黄俞淞 Photo by: Huang Yusong

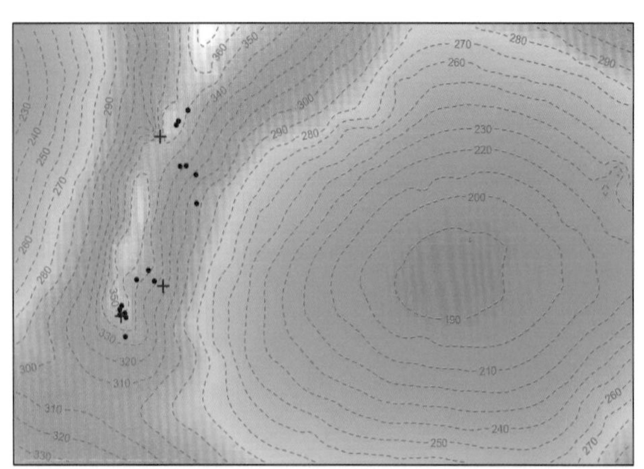

• 1~5 cm DBH + 5~20 cm DBH ○ ≥20 cm DBH
个体分布图 Distribution of individuals

径级分布表 DBH class

胸径区间 (Diameter class) (cm)	个体数 (No. of individuals in the plot)	比例 (Proportion) (%)
1~2	7	36.84
2~5	9	47.37
5~10	3	15.79
10~20	0	0.00
20~35	0	0.00
35~60	0	0.00
≥60	0	0.00

68 桂野桐

gùi yě tóng | Guangxi Mallotus

Mallotus conspurcatus Croiz.
大戟科 Euphorbiaceae

代码（SpCode） = MALCON
个体数（Individual number/15 hm^2） = 28
最大胸径（Max DBH） = 18.0 cm
重要值排序（Importance value rank） = 126

灌木，高约 3 m。小枝密被锈色星状长柔毛。叶革质，卵形或近圆形，顶端急尖或钝，基部圆形，上面无毛，下面密被褐红色星状长柔毛和黄色颗粒状腺体，掌状脉 5~8 条，叶柄密被红褐色星状毛。果序被柔毛，蒴果球形，直径约 1.5 cm，密生深棕色星状毛和软刺。果期 8~9 月。

Shrubs, ca. 3 m tall. Branchlets densely brown stellate-villous. Leaf blade ovate or nearly rounded, adaxially glabrous, abaxially densely brown stellate-villous and yellow granular gland, palmate vein 5–8, petiole densely red-brown stellate tomentose. Infructescence tomentose, capsule globose, ca. 1.5 cm in diam. densely brown stellate tomentose and softly spiny. Fr. Aug.–Sep..

植株 Whole plant
摄影：黄俞淞 Photo by：Huang Yusong

果序 Infructescence
摄影：黄俞淞 Photo by：Huang Yusong

叶 Leaf
摄影：黄俞淞 Photo by：Huang Yusong

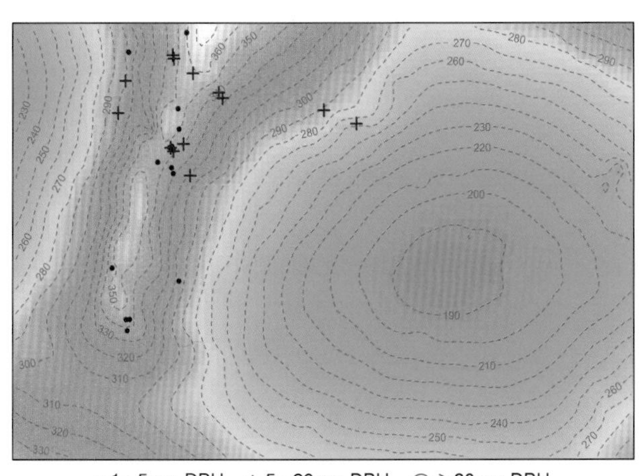

• 1~5 cm DBH + 5~20 cm DBH ○ ≥20 cm DBH
个体分布图 Distribution of individuals

径级分布表 DBH class

胸径区间 (Diameter class) (cm)	个体数 (No. of individuals in the plot)	比例 (Proportion) (%)
1~2	7	25.00
2~5	8	28.57
5~10	5	17.86
10~20	8	28.57
20~35	0	0.00
35~60	0	0.00
≥60	0	0.00

69 粗糠柴

cū kāng chái | Philippine Mallotus

Mallotus philippinensis (Lam.) Muell. Arg.
大戟科 Euphorbiaceae

代码（SpCode）= MALPHI
个体数（Individual number/15 hm^2）= 4
最大胸径（Max DBH）= 7.2 cm
重要值排序（Importance value rank）= 184

常绿小乔木，高 8~10 m。小枝被褐色星状柔毛。叶卵形至披针形，上面无毛，下面被稠密的短星状毛及红色腺点，基出 3 脉，近叶柄处有 2 腺体。花单性，雌雄同株，总状花序顶生或腋生。蒴果球形，密被鲜红色腺点及星状毛。花期 4~5 月，果期 5~8 月。

Evergreen small trees, 8–10 m tall. Branchlets brown stellate-villous. Leaf blade ovate to lanceolate, adaxially glabrous, abaxially densely stellate-tomentose and red gland, basal vein 3, with 2 gland near the apex of petiole. Flowers solitary, monecious, raceme terminal or axillary. Capsule globose, densely red gland and stellate-tomentose. Fl. Apr.–May,fr. May–Aug..

植株 Whole plant
摄影：黄俞淞 Photo by：Huang Yusong

果序 Infructescence
摄影：黄俞淞 Photo by：Huang Yusong

叶 Leaves
摄影：黄俞淞 Photo by：Huang Yusong

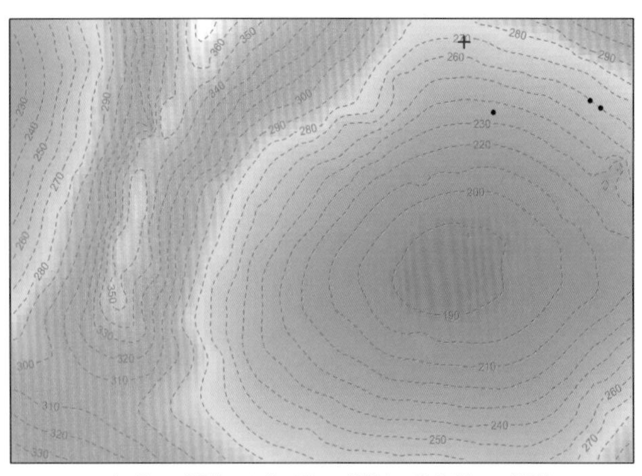

• 1~5 cm DBH + 5~20 cm DBH ○ ≥20 cm DBH
个体分布图 Distribution of individuals

径级分布表 DBH class

胸径区间 (Diameter class) (cm)	个体数 (No. of individuals in the plot)	比例 (Proportion) (%)
1~2	3	75.00
2~5	0	0.00
5~10	1	25.00
10~20	0	0.00
20~35	0	0.00
35~60	0	0.00
≥60	0	0.00

70 石岩枫

shí yán fēng | Repand Mallotus

Mallotus repandus (Willd.) Muell. Arg.
大戟科 Euphorbiaceae

代码（SpCode）= MALREP
个体数（Individual number/15 hm^2）= 26
最大胸径（Max DBH）= 7.6 cm
重要值排序（Importance value rank）= 139

攀缘状灌木。嫩枝、叶柄、花序和花梗均密生黄色星状柔毛。叶纸质或膜质，卵形或椭圆状卵形，基出脉 3 条，叶柄长 2~6 cm。花雌雄异株，雄花序顶生，稀腋生，雌花序顶生。蒴果具 2(3) 个分果爿，直径约 1 cm，密生黄色粉末状毛和具颗粒状腺体。花期 3~5 月，果期 8~9 月。

Climbing shrubs. Branchlets, petiole, inflorescence and pedicel densely yellow tomentose. Leaf blade papery or membranous, ovate or elliptic-ovate, basal vein 3, petiole 2–6 cm long. Flowers diecious, male inflorescence terminal, rarely axillary, female inflorescence terminal. Capsule 2(3)-locular, ca. 1 cm in diam., densely yellow pulverulent-tomentulose and granular gland. Fl. Mar.–May, fr. Aug.–Sep..

雌花序 Female inflorescence
摄影：黄俞淞 Photo by: Huang Yusong

果枝 Fruiting branch
摄影：黄俞淞 Photo by: Huang Yusong

雄花序 Male inflorescence
摄影：黄俞淞 Photo by: Huang Yusong

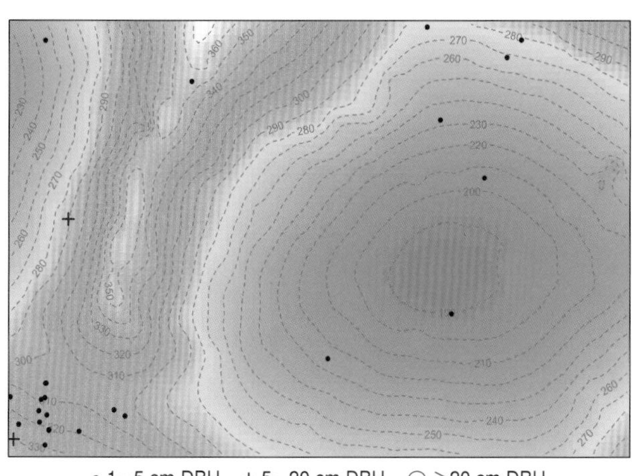

• 1~5 cm DBH + 5~20 cm DBH ○ ≥20 cm DBH
个体分布图 Distribution of individuals

径级分布表 DBH class

胸径区间 (Diameter class) (cm)	个体数 (No. of individuals in the plot)	比例 (Proportion) (%)
1~2	17	65.38
2~5	7	26.92
5~10	2	7.69
10~20	0	0.00
20~35	0	0.00
35~60	0	0.00
≥60	0	0.00

71 云南野桐

yún nán yě tóng | Yunnan Mallotus

Mallotus yunnanensis Pax. et K. Hoffmam
大戟科 Euphorbiaceae

代码（SpCode）= MALYUN
个体数（Individual number/15 hm^2）= 751
最大胸径（Max DBH）= 11.4 cm
重要值排序（Importance value rank）= 40

灌木，高 1.5~3 m。小枝和花序密被褐色星状短柔毛。叶对生，同对的叶大小不同，叶基具腺体 2 枚。花雌雄异株，雄花序总状，顶生或腋生；雌花序总状，顶生。蒴果扁球形，果表面具稀疏的短刺。花期 4~10 月，果期 10~12 月。

Shrubs, 1.5–3 m tall. Branchlets and inflorescence densely brownish stellate pubescent. Leaves opposite, each pair somewhat unequal, base with 2 glands. Flowers diecious, male raceme terminal or axillary, female raceme terminal. Capsule depressed globose, sparsely shortly softly spiny. Fl. Apr.–Oct., fr. Oct.–Dec..

果枝 Fruiting branch
摄影：黄俞淞 Photo by：Huang Yusong

果序 Infructescence
摄影：黄俞淞 Photo by：Huang Yusong

叶 Leaves
摄影：王斌 Photo by：Wang Bin

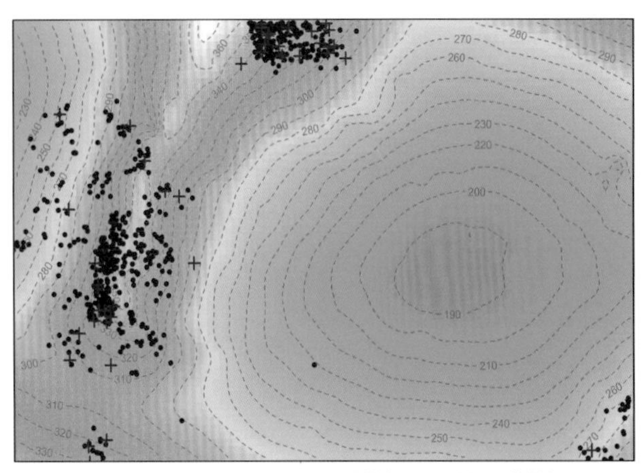

• 1~5 cm DBH + 5~20 cm DBH ○ ≥20 cm DBH
个体分布图 Distribution of individuals

径级分布表 DBH class

胸径区间 (Diameter class) (cm)	个体数 (No. of individuals in the plot)	比例 (Proportion) (%)
1~2	366	48.74
2~5	342	45.54
5~10	42	5.59
10~20	1	0.13
20~35	0	0.00
35~60	0	0.00
≥60	0	0.00

72 龙州珠子木

lóng zhōu zhū zǐ mù | Longzhou Phyllanthodendron

Phyllanthodendron breynioides P. T. Li
大戟科 Euphorbiaceae

代码（SpCode）= PHYBRE
个体数（Individual number/15 hm^2）= 14
最大胸径（Max DBH）= 3.8 cm
重要值排序（Importance value rank）= 163

灌木，高 1~3 m。小枝具棱。叶纸质，先端具小尖头，基部歪斜，叶柄被短柔毛。花雌雄同株，单生叶腋。蒴果球形，直径 1 cm。花期 7~8 月，果期 8~12 月。

Shrubs, 1–3 m tall. Branchlets angular. Leaf blade papery, apex with samll cusp, base oblique, petiole pubescent. Flowers monoecious, solitary, axillary. Capsules globose, 1 cm in diam.. Fl. Jul.–Aug, fr. Aug.–Dec..

植株 Whole plant
摄影：黄俞淞 Photo by：Huang Yusong

花枝 Flowering branch
摄影：黄俞淞 Photo by：Huang Yusong

果枝 Fruiting branch
摄影：黄俞淞 Photo by：Huang Yusong

• 1~5 cm DBH + 5~20 cm DBH ○ ≥20 cm DBH
个体分布图 Distribution of individuals

径级分布表 DBH class

胸径区间 (Diameter class) (cm)	个体数 (No. of individuals in the plot)	比例 (Proportion) (%)
1~2	11	78.57
2~5	3	21.43
5~10	0	0.00
10~20	0	0.00
20~35	0	0.00
35~60	0	0.00
≥60	0	0.00

73 余甘子

yú gān zǐ | Emblic

Phyllanthus emblica Linn.
大戟科 Euphorbiaceae

代码（SpCode）= PHYEMB
个体数（Individual number/15 hm^2）= 4
最大胸径（Max DBH）= 19.9 cm
重要值排序（Importance value rank）= 169

乔木，高可达 23 m。树皮褐色。叶片纸质至革质，二列，顶端截平或钝圆，有锐尖头或微凹，基部浅心形而稍偏斜。多朵雄花和 1 朵雌花或全为雄花组成腋生的聚伞花序。蒴果呈核果状，圆球形，直径 1~1.3 cm，外果皮肉质，内果皮硬壳质。花期 4~6 月，果期 7~9 月。

Trees, up to 23 m tall. Bark brownish. Leaf blade papery to leathery, distichous, apex truncate, or abtuse, mucronate or retuse, base shallowly cordate and slightly oblique. Fascicles with many male flowers and sometimes 1 or 2 larger female flowers. Capsule a drupe, globose, 1–1.3 cm in diam., exocarp fleshy, endocarp crustaceous. Fl. Apr.–Jun., fr. Jul.–Sep..

植株 Whole plant
摄影：黄俞淞 Photo by: Huang Yusong

果 Fruits
摄影：黄俞淞 Photo by: Huang Yusong

叶 Leaves
摄影：黄俞淞 Photo by: Huang Yusong

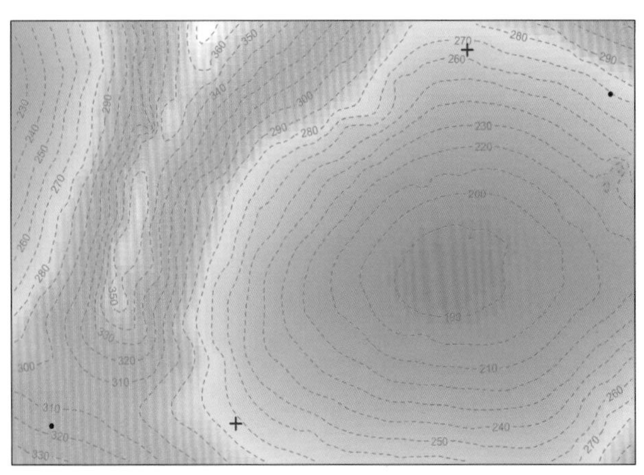

• 1~5 cm DBH + 5~20 cm DBH ○ ≥20 cm DBH
个体分布图 Distribution of individuals

径级分布表 DBH class

胸径区间 (Diameter class) (cm)	个体数 (No. of individuals in the plot)	比例 (Proportion) (%)
1~2	1	25.00
2~5	1	25.00
5~10	0	0.00
10~20	2	50.00
20~35	0	0.00
35~60	0	0.00
≥60	0	0.00

74 小果叶下珠　　xiǎo guǒ yè xià zhū | Reticulate Leafflower

Phyllanthus reticulatus Poir.
大戟科 Euphorbiaceae

代码（SpCode）= PHYRET
个体数（Individual number/15 hm^2）= 36
最大胸径（Max DBH）= 4.9 cm
重要值排序（Importance value rank）= 135

常绿灌木，高达 4 m。叶纸质，椭圆形至卵形，托叶软刺状。花雌雄同株，通常 1 朵雌花、5 朵雄花同生叶腋，雄花萼片 5 片，雄蕊 5 枚，其中 3 枚花丝合生成柱。果近球形，浆果状。花期 3~6 月，果期 6~10 月。

Evergreen shrubs, up to 4 m tall. Leaf blade papery, elliptic to ovate, stipule softly spiny. Flowers monoecious, axillary, with 1 male flower and 5 female flowers, female flowers: sepals 5, stamens 5, three of filament concrescent. Fruit a berry, subglobose. Fl. Mar.–Jun., fr. Jun.–Oct..

果枝　Fruiting branch
摄影：黄俞淞　Photo by：Huang Yusong

果　Fruits
摄影：黄俞淞　Photo by：Huang Yusong

枝叶　Branch and leaves
摄影：黄俞淞　Photo by：Huang Yusong

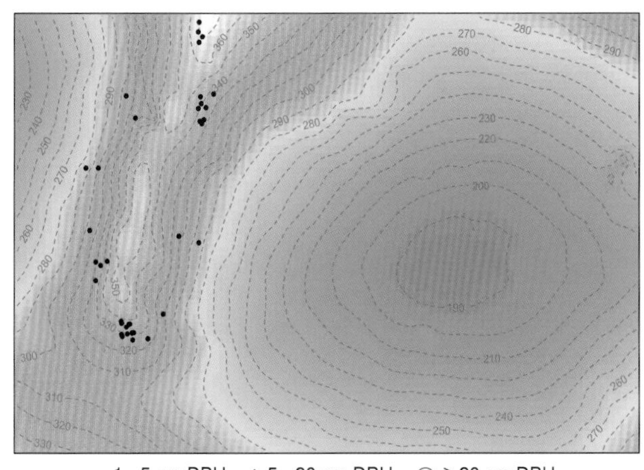

• 1~5 cm DBH　+ 5~20 cm DBH　○ ≥20 cm DBH
个体分布图　Distribution of individuals

径级分布表　DBH class

胸径区间 (Diameter class) (cm)	个体数 (No. of individuals in the plot)	比例 (Proportion) (%)
1~2	17	47.22
2~5	19	52.78
5~10	0	0.00
10~20	0	0.00
20~35	0	0.00
35~60	0	0.00
≥60	0	0.00

75 山乌桕

shān wū jiù | Mountain Chinese Tallow

Sapium discolor (Champ. ex Benth.) Muell. Arg.
大戟科 Euphorbiaceae

代码（SpCode）= SAPDIS
个体数（Individual number/15 hm^2）= 4
最大胸径（Max DBH）= 12.0 cm
重要值排序（Importance value rank）= 175

灌木或乔木，高 3~12 m。小枝灰褐色。叶互生，叶片椭圆形或长卵形，基部短狭或楔形，背面近缘常有数个圆形的腺体，叶柄长 2~7.5 cm，顶端具 2 毗连的腺体。花单性，雌雄同株，密集成长 4~9 cm 的顶生总状花序。蒴果黑色，球形。花期 4~6 月，果期 7~10 月。

Shrubs or trees, 3–12 m tall. Branchlets gray-brown. Leaves alternate, elliptic or oblong-ovate, base cuneate, with several rounded glands on or near margin abaxially, petioles 2–7.5 cm, 2-glandular at apex. Flowers monoecious in terminal racemes, inflorescences 4–9 cm. Capsules black, globose. Fl. Apr.–Jun., fr. Jul.–Oct..

果枝 Fruiting branch
摄影：黄俞淞 Photo by：Huang Yusong

果序 Infructescence
摄影：黄俞淞 Photo by：Huang Yusong

枝叶 Branch and leaves
摄影：黄俞淞 Photo by：Huang Yusong

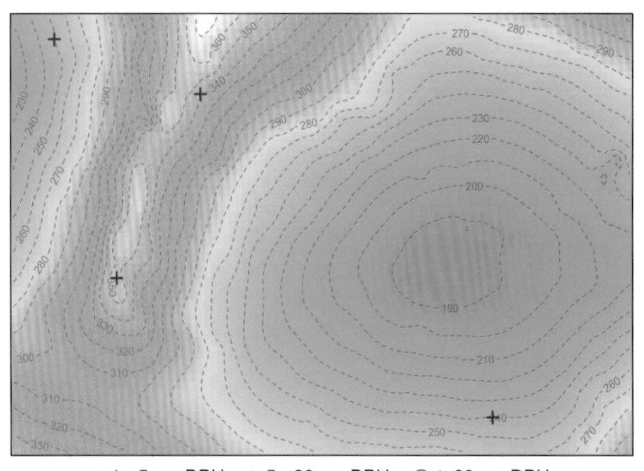

• 1~5 cm DBH + 5~20 cm DBH ○ ≥20 cm DBH
个体分布图 Distribution of individuals

径级分布表 DBH class

胸径区间 (Diameter class) (cm)	个体数 (No. of individuals in the plot)	比例 (Proportion) (%)
1~2	0	0.00
2~5	0	0.00
5~10	2	50.00
10~20	2	50.00
20~35	0	0.00
35~60	0	0.00
≥60	0	0.00

76 圆叶乌桕

yuán yè wū jiù | Round-leaf Triadica

Sapium rotundifolium Hemsl.
大戟科 Euphorbiaceae

代码（SpCode）= SAPROT
个体数（Individual number/15 hm^2）= 112
最大胸径（Max DBH）= 28.0 cm
重要值排序（Importance value rank）= 67

落叶灌木或乔木，高 3~12 m。幼枝常呈红色。叶薄革质，先端常具小凸尖，叶柄顶端具 2 腺体。花单性，雌雄同株，总状花序顶生。蒴果近球形。花期 4~6 月。

Deciduous shrubs or trees, 3–12 m tall. Branchlets usually red. Leaf blade slightly leathery, apex usually with cusp, 2 glands at the apex of petiole. Flowers solitary, monoecious, raceme terminal. Capsule subglobose. Fl. Apr.–Jun..

树干 Trunk
摄影：王斌 Photo by：Wang Bin

花枝 Flowering branch
摄影：黄俞淞 Photo by：Huang Yusong

果枝 Fruiting branch
摄影：黄俞淞 Photo by：Huang Yusong

• 1~5 cm DBH + 5~20 cm DBH ○ ≥20 cm DBH
个体分布图 Distribution of individuals

径级分布表 DBH class

胸径区间 (Diameter class) (cm)	个体数 (No. of individuals in the plot)	比例 (Proportion) (%)
1~2	5	4.46
2~5	15	13.39
5~10	25	22.32
10~20	58	51.79
20~35	9	8.04
35~60	0	0.00
≥60	0	0.00

77 宿萼木

sù è mù | Fringe-calyx Strophioblachia

Strophioblachia fimbricalyx Boerl.
大戟科 Euphorbiaceae

代码（SpCode）= STRFIM
个体数（Individual number/15 hm^2）= 15
最大胸径（Max DBH）= 29.6 cm
重要值排序（Importance value rank）= 154

常绿灌木，高 2~4 m。叶膜质，卵状披针形或长圆状倒卵形。总状花序聚伞状，萼片约 8 mm，花后增大至 18 mm，边缘密生长约 5 mm 的腺毛，无花瓣。蒴果钝三棱状球形，基部具宿存萼片。花期 5~7 月。

Evergreen shrubs, 2–4 m tall. Leaf blade membranous, ovate-lanceolate or oblong-oblanceolate. Raceme cymose, sepals ca. 8 mm long, increase to 18 mm long when flowered, margin with dense ca. 5 mm glandular hairs, petals absent. Capsules ovoid-globose, slightly depressed, base with persistent calyx. Fl. May–Jul..

果枝 Fruiting branches
摄影：黄俞淞 Photo by：Huang Yusong

花 Flower
摄影：黄俞淞 Photo by：Huang Yusong

叶 Leaves
摄影：黄俞淞 Photo by：Huang Yusong

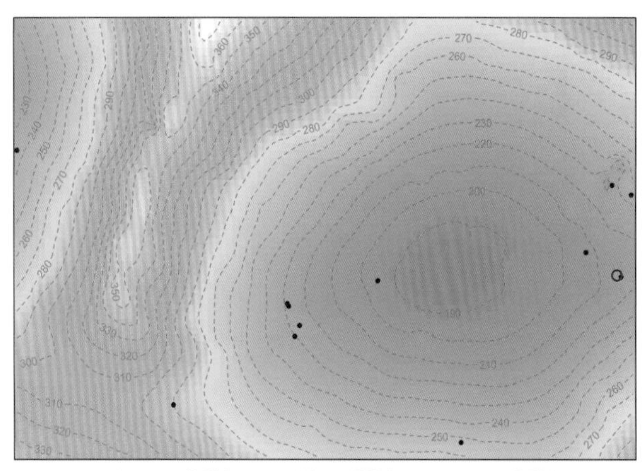
• 1~5 cm DBH + 5~20 cm DBH ○ ≥20 cm DBH
个体分布图 Distribution of individuals

径级分布表 DBH class

胸径区间 (Diameter class) (cm)	个体数 (No. of individuals in the plot)	比例 (Proportion) (%)
1~2	12	80.00
2~5	2	13.33
5~10	0	0.00
10~20	0	0.00
20~35	1	6.67
35~60	0	0.00
≥60	0	0.00

78 孟仑三宝木 mèng lún sān bǎo mù | Bon's Trigonostemon

Trigonostemon bonianus Gagnep.
大戟科 Euphorbiaceae

代码（SpCode）= TRIBON
个体数（Individual number/15 hm^2）= 14
最大胸径（Max DBH）= 4.0 cm
重要值排序（Importance value rank）= 157

常绿灌木或小乔木，高 1.5~4 m。叶纸质，卵形、椭圆形至长圆状披针形，两面无毛，基出脉 3 条。聚伞花序生于枝条近顶端；蒴果近球形，无毛。花期 3~8 月。

Evergreen shrubs or small trees, ca. 1.5–4 m tall. Leaf blade papery, ovate, elliptic to oblong-lanceolate, glabrous on both surfaces, basal veins 3. Cyme nearly terminal. Capsule subglobose, glabrous. Fl. Mar.–Aug..

植株 Whole plant
摄影：黄俞淞 Photo by：Huang Yusong

花枝 Flowering branch
摄影：黄俞淞 Photo by：Huang Yusong

果 Fruits
摄影：刘晟源 Photo by：Liu Shengyuan

• 1~5 cm DBH + 5~20 cm DBH ○ ≥20 cm DBH
个体分布图 Distribution of individuals

径级分布表 DBH class

胸径区间 (Diameter class) (cm)	个体数 (No. of individuals in the plot)	比例 (Proportion) (%)
1~2	12	85.71
2~5	2	14.29
5~10	0	0.00
10~20	0	0.00
20~35	0	0.00
35~60	0	0.00
≥60	0	0.00

79 大花枇杷

dà huā pí pá | Bigflower Loquat

Eriobotrya cavaleriei (H. Lév.) Rehder
蔷薇科 Rosaceae

代码（SpCode）= ERICAV
个体数（Individual number/15 hm^2）= 1
最大胸径（Max DBH）= 1.0 cm
重要值排序（Importance value rank）= 217

常绿乔木，高 4~6 m。小枝粗壮，无毛。叶长圆形、长圆状披针形或长圆状倒披针形。圆锥花序顶生，总花梗和花梗有稀疏棕色短柔毛；果实椭圆形或近球形，肉质。花期 4~5 月，果期 7~8 月。

Evergreen trees, 4–6 m tall. Branchlets stout, glabrous. Leaf blade oblong, oblong-lanceolate or oblong-oblanceolate. Panicle terminal, peduncle sparsely brown pubescent. Fruit elliptic or subglobose, fleshy. Fl. Apr.–May, fr. Jul.–Aug..

枝叶 Branch and leaves
摄影：黄俞淞 Photo by：Huang Yusong

枝 Branch
摄影：黄俞淞 Photo by：Huang Yusong

叶 Leaves
摄影：黄俞淞 Photo by：Huang Yusong

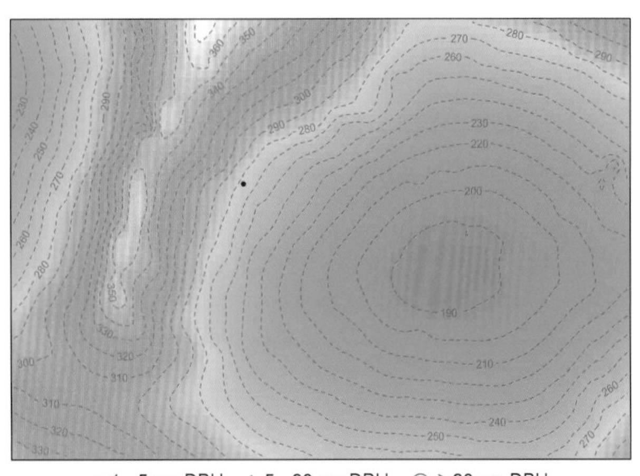

• 1~5 cm DBH　+ 5~20 cm DBH　○ ≥20 cm DBH
个体分布图 Distribution of individuals

径级分布表 DBH class

胸径区间 (Diameter class) (cm)	个体数 (No. of individuals in the plot)	比例 (Proportion) (%)
1~2	1	100.00
2~5	0	0.00
5~10	0	0.00
10~20	0	0.00
20~35	0	0.00
35~60	0	0.00
≥60	0	0.00

80 大叶桂樱

dà yè guì yīng | Bigleaf Cherrylaurel

Laurocerasus zippeliana (Miq.) T.T. Yu L.T. Lu
蔷薇科 Rosaceae

代码（SpCode）= LAUZIP
个体数（Individual number/15 hm^2）= 275
最大胸径（Max DBH）= 26.9 cm
重要值排序（Importance value rank）= 35

常绿乔木，高达 20 m。小枝紫褐色或黑褐色。叶革质，基部宽楔形至近圆形，叶边具稀疏或稍密粗锯齿，或具针刺状齿。总状花序单生或 2~4 个簇生于叶腋；果实长圆形或卵状长圆形，褐色或黑褐色。花期 7~10 月，果期冬季。

Evergreen trees, up to 20 m tall. Branchlets purplish brown or blackish brown. Leaf blade leathery, base broadly cuneate to sub rounded, margin with sparse to dense coarse black glandular serration, or with a few acicular teeth. Racemes solitary or 2–4 in a fascicle axillary. Fruit oblong to ovoid-oblong, brown to blackish brown. Fl. Jul.–Oct., fr. winter.

树干 Trunk
摄影：黄俞淞 Photo by：Huang Yusong

叶 Leaf
摄影：王斌 Photo by：Wang Bin

枝叶 Branch and leaves
摄影：黄俞淞 Photo by：Huang Yusong

• 1~5 cm DBH + 5~20 cm DBH ○ ≥20 cm DBH
个体分布图 Distribution of individuals

径级分布表 DBH class

胸径区间 (Diameter class) (cm)	个体数 (No. of individuals in the plot)	比例 (Proportion) (%)
1~2	38	13.82
2~5	106	38.55
5~10	63	22.91
10~20	62	22.55
20~35	6	2.18
35~60	0	0.00
≥60	0	0.00

81 臀果木

tún guǒ mù | Topeng's Pygeum

Pygeum topengii Merr.
蔷薇科 Rosaceae

代码（SpCode）= PYGTOP
个体数（Individual number/15 hm^2）= 23
最大胸径（Max DBH）= 48.0 cm
重要值排序（Importance value rank）= 115

乔木，高可达 25 m。树皮灰褐色, 小枝暗褐色。叶片革质，卵状椭圆形或椭圆形，全缘，下面被平铺褐色柔毛，近基部有 2 枚黑色腺体，叶柄长 5~8 mm，被褐色柔毛。总状花序，单生或数个簇生于叶腋。果实肾形，顶端常无突尖而凹陷，无毛。花期 6~9 月，果期冬季。

Trees, up to 25 m tall. Bark dark gray, branchlets dark brown. Leaf blade leathery, ovate-elliptic, entire, abaxially brown appressed pubescent and with 2 black glands near base, petiole 5–8 mm long, brown pubescent. Racemes solitary or to several in a fascicle, axillary. Fruit reniform, apically depressed, glabrous. Fl. Jun.–Sep., fr. winter.

树干 Trunk
摄影：黄俞淞 Photo by：Huang Yusong

枝叶 Branch and leaves
摄影：黄俞淞 Photo by：Huang Yusong

果 Fruits
摄影：黄俞淞 Photo by：Huang Yusong

• 1~5 cm DBH + 5~20 cm DBH ○ ≥20 cm DBH
个体分布图 Distribution of individuals

径级分布表 DBH class

胸径区间 (Diameter class) (cm)	个体数 (No. of individuals in the plot)	比例 (Proportion) (%)
1~2	5	21.74
2~5	7	30.43
5~10	4	17.39
10~20	2	8.70
20~35	4	17.39
35~60	1	4.35
≥60	0	0.00

82 海红豆

hǎi hóng dòu | Red Beadtree

Adenanthera pavonina L.
含羞草科 Mimosacea

代码（SpCode）= ADEPAV
个体数（Individual number/15 hm^2）= 94
最大胸径（Max DBH）= 33.6 cm
重要值排序（Importance value rank）= 80

落叶乔木，高 5~20 m。嫩枝被柔毛。二回羽状复叶，羽片 3~5 对，小叶 4~7 对，互生，长圆形或卵形，两端圆钝。总状花序单生于叶腋或在枝顶排成圆锥花序，被短柔毛。荚果长圆形，盘旋，开裂后果瓣旋卷，种子鲜红色。花期 4~7 月，果期 7~10 月。

Deciduous trees, 5–20 m tall. Branchlets pubescent. Bipinnate leaf, pinnae 3–5 pairs, leaflets 4–7 pairs, alternate, oblong or ovate, both ends obtruse. Racemes axillary, solitary, or heads arranged in panicles, pubescent. Legume oblong, spiral, carpel convolute when cracked, seeds red. Fl. Apr.–Jul., fr. Jul.–Oct..

树干 Trunk
摄影：丁涛 Photo by: Ding Tao

荚果 Legume fruits
摄影：黄俞淞 Photo by: Huang Yusong

复叶 Compound leaves
摄影：黄俞淞 Photo by: Huang Yusong

• 1~5 cm DBH + 5~20 cm DBH ○ ≥20 cm DBH
个体分布图 Distribution of individuals

径级分布表 DBH class

胸径区间 (Diameter class) (cm)	个体数 (No. of individuals in the plot)	比例 (Proportion) (%)
1~2	29	30.85
2~5	28	29.79
5~10	15	15.96
10~20	18	19.15
20~35	4	4.26
35~60	0	0.00
≥60	0	0.00

83 香合欢

xiāng hé huān | Ceylon Rosewood

Albizia odoratissima (Linn. f.) Benth.
含羞草科 Mimosacea

代码（SpCode）= ALBODO
个体数（Individual number/15 hm^2）= 49
最大胸径（Max DBH）= 38.6 cm
重要值排序（Importance value rank）= 97

常绿乔木，高 5~15 m。小枝初被柔毛。二回羽状复叶，羽片 2~4(6) 对；小叶 6~14 对，纸质，长圆形，两面稍被贴生、稀疏短柔毛，中脉偏于上缘，无柄。头状花序排成顶生、疏散的圆锥花序，被短柔毛。荚果长圆形，扁平。花期 4~7 月，果期 6~10 月。

Evergreen trees , 5–15 m tall. Branchlets pubescent when young. Bipinnate leaf, pinnae 2–4(6) pairs, papery, oblong, both surfaces sparsely appressed pubescent, main vein close to upper margin, without petiole. Heads arranged in panicles, puberulent to tomentose. Legume oblong, appressed. Fl. Apr.–Jul., fr. Jun.–Oct..

树干 Trunk
摄影：王斌 Photo by: Huang Yusong

花序 Inflorescence
摄影：黄俞淞 Photo by: Huang Yusong

复叶 Compound leaves
摄影：黄俞淞 Photo by: Huang Yusong

• 1~5 cm DBH + 5~20 cm DBH ○ ≥20 cm DBH
个体分布图 Distribution of individuals

径级分布表 DBH class

胸径区间 (Diameter class) (cm)	个体数 (No. of individuals in the plot)	比例 (Proportion) (%)
1~2	12	24.49
2~5	15	30.61
5~10	8	16.33
10~20	9	18.37
20~35	3	6.12
35~60	2	4.08
≥60	0	0.00

84 广西棋子豆

guǎng xī qí zǐ dòu | Guangxi Archidendron

Archidendron guangxiensis T. L. Wu
含羞草科 Mimosacea

代码（SpCode）= ARCGUA
个体数（Individual number/15 hm^2）= 447
最大胸径（Max DBH）= 45.7 cm
重要值排序（Importance value rank）= 25

常绿乔木，高可达 15 m。二回羽状复叶，羽片 2 对，小叶片 2~3 对，最下面的小叶片偶为 1 片, 总叶柄着生 1 枚腺体，顶端 1 对小羽片叶柄间具 1 枚腺体。花期 3~5 月，果期 7~8 月。

Evergreen trees, up to 15 m tall. Bipinnate leaf, pinnae 2 pairs, leaflets 2–3 pairs, common petiole with 1 gland, between the petioles of terminal leaflets with 1 gland. Fl. Mar.–May, fr. Jul.–Aug..

树干 Trunk
摄影：王斌 Photo by：Wang Bin

复叶 Compound leaf
摄影：王斌 Photo by：Wang Bin

种子 Seeds
摄影：王斌 Photo by：Wang Bin

• 1~5 cm DBH + 5~20 cm DBH ○ ≥20 cm DBH

个体分布图 Distribution of individuals

径级分布表 DBH class

胸径区间 (Diameter class) (cm)	个体数 (No. of individuals in the plot)	比例 (Proportion) (%)
1~2	167	37.36
2~5	101	22.60
5~10	52	11.63
10~20	55	12.30
20~35	67	14.99
35~60	5	1.12
≥60	0	0.00

85 顶果木

dǐng guǒ mù | Kuranjan

Acrocarpus fraxinifolius Wight ex Arn.
含羞草科 Mimosacea

代码（SpCode）= ACRFRA
个体数（Individual number/15 hm^2）= 1
最大胸径（Max DBH）= 20.7 cm
重要值排序（Importance value rank）= 195

高大乔木，枝下高可达 30 m。二回羽状复叶长 30~40 cm，下部的叶具羽片 3~8 对，小叶 4~8 对，对生，近革质，卵形或卵状长圆形，全缘。总状花序腋生，具密集的花。荚果扁平，沿腹缝线具狭翅，翅宽 3~5 mm；种子 14~18 颗，淡褐色。花期 3~4 月，果期 6~7 月。

Trees, large, to 30 m tall before tree forks. Bipinnate leaf, 30–40 cm long, lower leaves with 3–8 pairs pinnae, leaflets 4–8 pairs, nearly leathery, ovate or ovate-oblong, margin entire. Racemes axillary, densely flowers. Legume appressed, narrowly winged along ventral suture, wing 3–5 mm wide, seeds 14–18, light brown. Fl. Mar.–Apr., fr. Jun.–Jul..

植株 Whole plant
摄影：黄俞淞 Photo by: Huang Yusong

荚果 Legume fruits
摄影：黄俞淞 Photo by: Huang Yusong

复叶 Compound leaf
摄影：丁涛 Photo by: Ding Tao

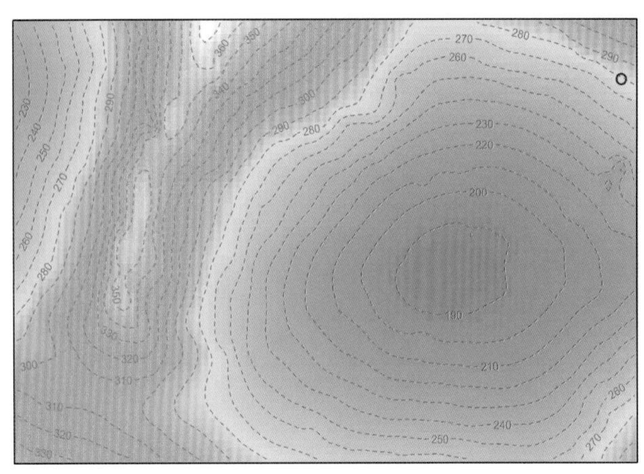

• 1~5 cm DBH　+ 5~20 cm DBH　○ ≥20 cm DBH
个体分布图 Distribution of individuals

径级分布表 DBH class

胸径区间 (Diameter class) (cm)	个体数 (No. of individuals in the plot)	比例 (Proportion) (%)
1~2	0	0.00
2~5	0	0.00
5~10	0	0.00
10~20	0	0.00
20~35	1	100.00
35~60	0	0.00
≥60	0	0.00

86 华南皂荚

huá nán zào jiǎ | South China Honeylocust

Gleditsia fera (Lour.) Merr.
云实科 Caesalpiniaceae

代码（SpCode）= GLEFER
个体数（Individual number/15 hm^2）= 65
最大胸径（Max DBH）= 29.0 cm
重要值排序（Importance value rank）= 102

常绿小乔木至乔木，高 3~12 m。枝灰褐色。叶为一回羽状复叶，边缘具圆齿，中脉在小叶基部偏斜。花数朵组成小聚伞花序，再由多个聚伞花序组成腋生或顶生、长 7~16 cm 的总状花序。荚果扁平，先端具 2~5 mm 的喙。花期 4~5 月，果期 6~12 月。

Evergreen small trees to trees, 3–12 m tall. Bark gray brown. Leaves pinnate, margin crenate, midvein inclined to one side at base of blade. Flowers several in cymules, axillary or terminal racemes 7–16 cm. Legume appressed, apex with beak 2–5 mm. Fl. Apr.–May, fr. Jun.–Dec..

树干 Trunk
摄影：黄俞淞 Photo by：Huang Yusong

复叶 Compound leaves
摄影：黄俞淞 Photo by：Huang Yusong

枝叶 Branch and leaves
摄影：黄俞淞 Photo by：Huang Yusong

• 1~5 cm DBH + 5~20 cm DBH ○ ≥20 cm DBH
个体分布图 Distribution of individuals

径级分布表 DBH class

胸径区间 (Diameter class) (cm)	个体数 (No. of individuals in the plot)	比例 (Proportion) (%)
1~2	18	27.69
2~5	26	40.00
5~10	13	20.00
10~20	5	7.69
20~35	3	4.62
35~60	0	0.00
≥60	0	0.00

87 小果皂荚

xiǎo guǒ zào jiá | Small-fruit Honeylocust

Gleditsia australis Hemsl.
云实科 Caesalpiniaceae

代码（SpCode）= GLEAUS
个体数（Individual number/15 hm^2）= 13
最大胸径（Max DBH）= 11.1 cm
重要值排序（Importance value rank）= 162

小乔木至乔木，高 3~20 m。枝褐灰色，具粗刺，刺圆锥状，有分枝。叶为一回或二回羽状复叶（具羽片 2~6 对），纸质至薄革质，边缘具钝齿或近全缘。雄花数朵簇生，组成总状花序，再组成圆锥花序，腋生或顶生。荚果带状长圆形，压扁。花期 6~10 月；果期 11 月至翌年 4 月。

Small trees to trees, 3–20 m tall. Branches brownish gray, with robust spines, spines conical, branched. Leaves pinnate or bipinnate, pinnae 2–6 pairs, papery to thinly leathery, margin obtuse or nearly entire. Male flowers several fascicled in dense racemes, several racemes comprising a panicle, axillary or terminal. Legume strap-shaped, compressed. Fl. Jun.–Oct., fr. Nov.–Apr. of next year.

树干 Trunk
摄影：王斌 Photo by：Wang Bin

复叶 Compound leaves
摄影：黄俞淞 Photo by：Huang Yusong

枝刺 Branched spines
摄影：王斌 Photo by：Wang Bin

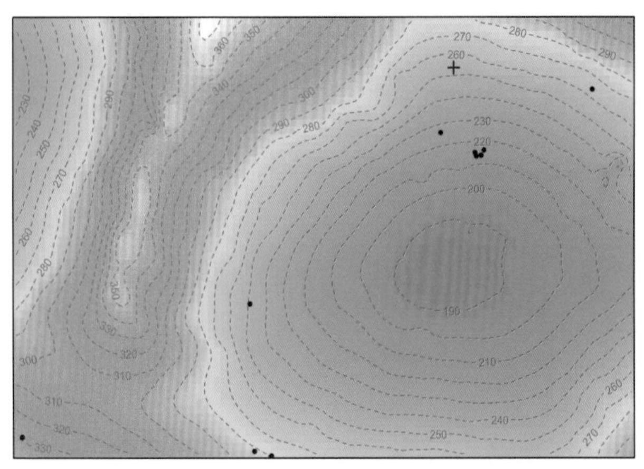

• 1~5 cm DBH + 5~20 cm DBH ○ ≥20 cm DBH
个体分布图 Distribution of individuals

径级分布表 DBH class

胸径区间 (Diameter class) (cm)	个体数 (No. of individuals in the plot)	比例 (Proportion) (%)
1~2	8	61.54
2~5	4	30.77
5~10	0	0.00
10~20	1	7.69
20~35	0	0.00
35~60	0	0.00
≥60	0	0.00

88 中国无忧花

zhōng guó wú yōu huā | Chinese Saraca

Saraca dives Pierre
云实科 Caesalpiniaceae

代码（SpCode）＝ SARDIV
个体数（Individual number/15 hm^2）＝ 162
最大胸径（Max DBH）＝ 75.1 cm
重要值排序（Importance value rank）＝ 31

常绿乔木，高 5~20 m。羽状复叶，有小叶 5~6 对，长圆形、卵状披针形或长倒卵形。花序腋生；荚果棕褐色，开裂后果瓣卷曲。花期 4~5 月，果期 7~10 月。

Evergreen trees, 5–20 m tall. Pinnately leaf, leaflets 5–6 pairs, oblong, ovate-lanceolate ot obovate. Inflorescence axillary. Legume brownish, carpel convolute when cracked. Fl. Apr.–May, fr. Jul.–Oct..

树干 Trunk
摄影：王斌 Photo by：Wang Bin

嫩枝 New branches
摄影：丁涛 Photo by：Ding Tao

花 Flowers
摄影：刘晟源 Photo by：Liu Shengyuan

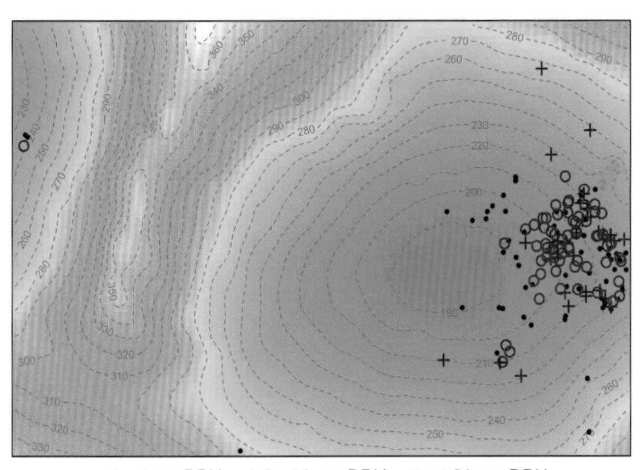

• 1~5 cm DBH + 5~20 cm DBH ○ ≥20 cm DBH

个体分布图 Distribution of individuals

径级分布表 DBH class

胸径区间 (Diameter class) (cm)	个体数 (No. of individuals in the plot)	比例 (Proportion) (%)
1~2	30	18.52
2~5	36	22.22
5~10	13	8.02
10~20	18	11.11
20~35	50	30.86
35~60	12	7.41
≥60	3	1.85

89 任豆

rén dòu | Common Zenia

Zenia insignis Chun
云实科 Caesalpiniaceae

代码（SpCode）= ZENINS
个体数（Individual number/15 hm^2）= 1
最大胸径（Max DBH）= 21.8 cm
重要值排序（Importance value rank）= 193

乔木，高 15~20 m。小枝黑褐色，散生有黄白色的小皮孔。叶长 25~45 cm；小叶薄革质，长圆状披针形，边全缘，下面有灰白色的糙伏毛。圆锥花序顶生，总花梗和花梗被黄色或棕色糙伏毛。荚果长圆形或椭圆状长圆形，红棕色。花期 5 月，果期 6~8 月。

Trees, 15–20 m tall. Branchlets blackish brown, with scattered, yellowish white, small lenticels. Leaves 25–45 cm long, leaflets thinly leathery, oblong-lanceolate, margin entire, abaxially grayish white strigose. Panicles terminal, peduncles and pedicels yellow or brown strigose. Legume oblong or elliptic-oblong, reddish brown. Fl. May, fr. Jun.–Aug..

枝叶 Branch and leaves
摄影：黄俞淞 Photo by: Huang Yusong

果枝 Fruiting branch
摄影：黄俞淞 Photo by: Huang Yusong

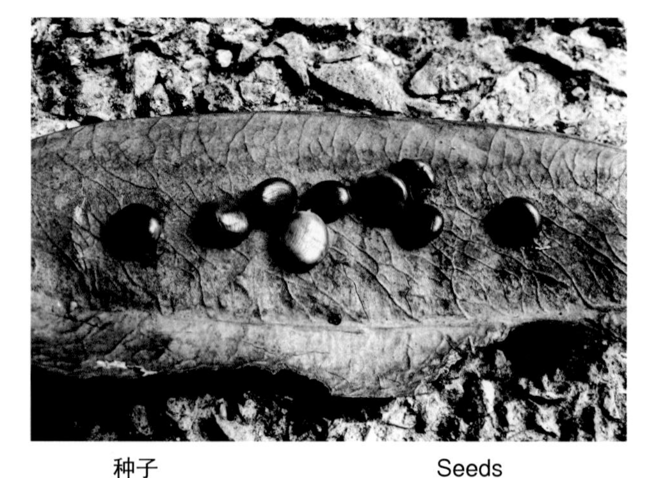

种子 Seeds
摄影：黄俞淞 Photo by: Huang Yusong

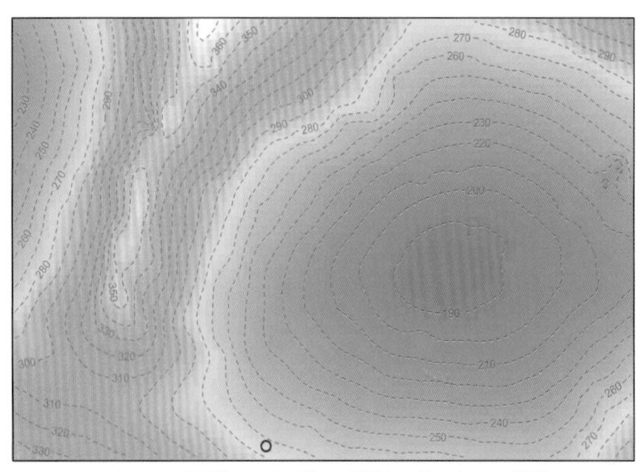

• 1~5 cm DBH + 5~20 cm DBH ○ ≥20 cm DBH
个体分布图 Distribution of individuals

径级分布表 DBH class

胸径区间 (Diameter class) (cm)	个体数 (No. of individuals in the plot)	比例 (Proportion) (%)
1~2	0	0.00
2~5	0	0.00
5~10	0	0.00
10~20	0	0.00
20~35	1	100.00
35~60	0	0.00
≥60	0	0.00

90 小托叶密脉杭子梢 xiǎo tuō yè mì mài háng zǐ shāo | Stipellate Clovershrub

Campylotropis bonii Schindl. var. *stipellata* Iokawa et H. Ohashi
蝶形花科 Fabaceae

代码（SpCode）= CAMBON
个体数（Individual number/15 hm^2）= 1
最大胸径（Max DBH）= 1.1 cm
重要值排序（Importance value rank）= 210

落叶灌木，高 1~3 m。小枝有细棱。羽状复叶具 3 小叶，先端微凹至圆形，具小凸尖。总状花序通常单一腋生并顶生。荚果压扁，先端喙尖长 0.3~0.8 mm，表面具短绢毛。花期 10~12 月，果期 11~12 月。

Deciduous shrubs, 1–3 m tall. Twig angulate. Leaves pinnately, leaflets 3, apex retuse to rounded, with small cusp. Raceme solitary, axillary or terminal. Legume compressed, beak 0.3–0.8 mm long at apex, surfaces with short silky hair. Fl. Oct.–Dec., fr. Nov.–Dec..

花枝 Flowering branches
摄影：黄俞淞 Photo by：Huang Yusong

花序 Inflorescence
摄影：黄俞淞 Photo by：Huang Yusong

复叶 Compound leaves
摄影：黄俞淞 Photo by：Huang Yusong

• 1~5 cm DBH + 5~20 cm DBH ○ ≥20 cm DBH
个体分布图 Distribution of individuals

径级分布表 DBH class

胸径区间 (Diameter class) (cm)	个体数 (No. of individuals in the plot)	比例 (Proportion) (%)
1~2	1	100.00
2~5	0	0.00
5~10	0	0.00
10~20	0	0.00
20~35	0	0.00
35~60	0	0.00
≥60	0	0.00

91 翅荚香槐

chì jiá xiāng huái | Broad-fruit Yellowwood

Cladrastis platycarpa (Maxim.) Makino
蝶形花科 Fabaceae

代码（SpCode）= CLAPLA
个体数（Individual number/15 hm^2）= 1
最大胸径（Max DBH）= 3.0 cm
重要值排序（Importance value rank）= 211

乔木，高 30 m。树皮暗灰色，多皮孔。奇数羽状复叶；小叶 3~4 对，互生或近对生，长椭圆形或卵状长圆形，基部的最小，顶生的最大。圆锥花序长 30 cm，花序轴和花梗被疏短柔毛。荚果扁平，长椭圆形或长圆形，两侧具翅，不开裂。花期 4~6 月，果期 7~10 月。

Trees, 30 m tall. Bark dark gray, many lenticellate. Leaves imparipinnately compound, leaflets 3–4 pairs, alternate or nearly opposite, oblong-elliptic or ovate-oblong, base minimum, apex maximum. Panicle 30 cm long, peduncles and pedicels pubescent. Legume compressed, oblong-elliptic or oblong, winged on both sides, indehiscent. Fl. Apr.–Jun., fr. Jul.–Oct..

枝叶 Branch and leaves
摄影：黄俞淞 Photo by：Huang Yusong

果枝 Fruiting branch
摄影：黄俞淞 Photo by：Huang Yusong

果序 Infructescence
摄影：黄俞淞 Photo by：Huang Yusong

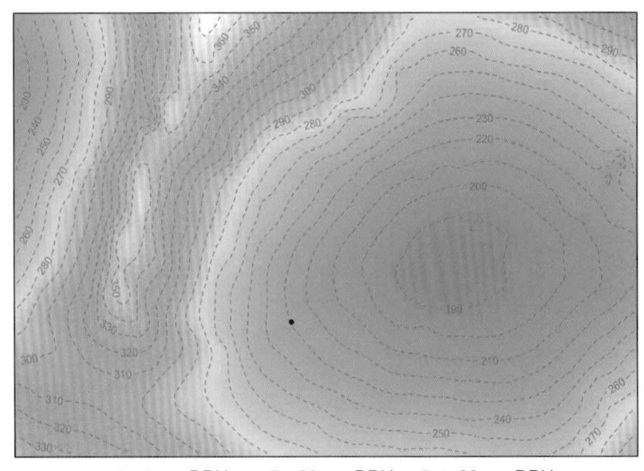

• 1~5 cm DBH + 5~20 cm DBH ○ ≥20 cm DBH
个体分布图 Distribution of individuals

径级分布表 DBH class

胸径区间 (Diameter class) (cm)	个体数 (No. of individuals in the plot)	比例 (Proportion) (%)
1~2	0	0.00
2~5	1	100.00
5~10	0	0.00
10~20	0	0.00
20~35	0	0.00
35~60	0	0.00
≥60	0	0.00

92 假木豆

jiǎ mù dòu | Trigonous-branch Dendrolobium

Dendrolobium triangulare (Retz.) Schindl.
蝶形花科 Fabaceae

代码（SpCode）= DENTRI
个体数（Individual number/15 hm^2）= 1
最大胸径（Max DBH）= 1.4 cm
重要值排序（Importance value rank）= 209

常绿灌木，高 1~2 m。叶为三出羽状复叶，上面无毛，下面被长丝状毛。花序腋生，伞形花序有花 20~30 朵；荚果长 2~2.5 cm，稍弯曲，有荚节 3~6 节，被贴伏丝状毛。花期 8~10 月，果期 10~12 月。

Evergreen shrubs, 1–2 m tall. Leaves 3-foliolate, adaxially glabrous, abaxially long sericeous especially on veins. Inflorecence axillary, umbels 20–30-flowered. Legume 2–2.5 cm long, slightly arcuate, 3–6-jointed, appressed sericeous. Fl. Aug.–Oct., fr. Oct.–Dec..

嫩叶 New Leaves
摄影：黄俞淞 Photo by：Huang Yusong

花枝 Flowering branch
摄影：黄俞淞 Photo by：Huang Yusong

荚果 Legumes
摄影：黄俞淞 Photo by：Huang Yusong

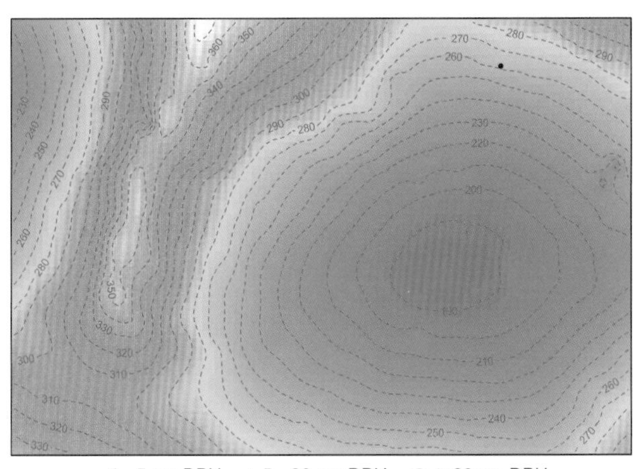

• 1~5 cm DBH + 5~20 cm DBH ○ ≥20 cm DBH
个体分布图 Distribution of individuals

径级分布表 DBH class

胸径区间 (Diameter class) (cm)	个体数 (No. of individuals in the plot)	比例 (Proportion) (%)
1~2	1	100.00
2~5	0	0.00
5~10	0	0.00
10~20	0	0.00
20~35	0	0.00
35~60	0	0.00
≥60	0	0.00

93 劲直刺桐

jìng zhí cì tóng | Straight Coralbean

Erythrina stricta Roxb.
蝶形花科 Fabaceae

代码（SpCode）= ERYSTR
个体数（Individual number/15 hm^2）= 316
最大胸径（Max DBH）= 64.2 cm
重要值排序（Importance value rank）= 6

落叶乔木，高 7~12 m。树干通直，小枝具短、圆锥状的刺。叶为羽状 3 小叶，两面无毛。总状花序长 15 cm，花 3 朵一束，鲜红色。荚果通常有种子 1~3 粒。花期 3 月，果期 8 月。

Deciduous trees, 7–12 m tall. Trunk straight, branches with short conic prickles. Leaves pinnately 3-foliolate, both surfaces glabrous. Raceme 15 cm long, 3-clustered, scarlet. Legume usually with 1–3 seeds. Fl. Mar., fr. Aug..

树干 Trunk
摄影：王斌 Photo by：Wang Bin

复叶 Compound leaf
摄影：黄俞淞 Photo by：Huang Yusong

花 Flowers
摄影：黄俞淞 Photo by：Huang Yusong

• 1~5 cm DBH + 5~20 cm DBH ○ ≥20 cm DBH
个体分布图 Distribution of individuals

径级分布表 DBH class

胸径区间 (Diameter class) (cm)	个体数 (No. of individuals in the plot)	比例 (Proportion) (%)
1~2	10	3.16
2~5	13	4.11
5~10	15	4.75
10~20	79	25.00
20~35	148	46.84
35~60	50	15.82
≥60	1	0.32

94 槟榔柯

bīng láng kē | Betelnut-palm Tanoak

Lithocarpus areca (Hickel et A. Camus) A. Camus
壳斗科 Fagaceae

代码（SpCode）＝ LITARE
个体数（Individual number/15 hm^2）＝ 7
最大胸径（Max DBH）＝ 6.0 cm
重要值排序（Importance value rank）＝ 177

常绿乔木，高 10~20 m。小枝灰白色。叶倒披针形，叶缘上部有少数浅裂的锐齿或全缘。雄穗状花序单生腋生，雌花序通常雌雄同序，果序粗短，通常一簇中仅一个发育；坚果椭圆形，或长圆锥形，上部有 3 条纵向略呈钝角的脊棱。花期 10 月，果翌年 11 月成熟。

Evergreen trees, 10–20 m tall. Branchlets grayish white. Leaf blade oblanceolate, margin with a few sharp teeth from middle to apex or sometimes entire. Male inflorescent axillary, solitary, female inflorescent often androgynous with female flowers on basal part of rachis, usually 1 developed. Nut elliptic, or long conical, with 3 longitudinal obtuse ridges. Fl. Oct., fr. ripe in Nov. of following year.

枝叶 Branch and leaves
摄影：黄俞淞 Photo by：Huang Yusong

果 Fruits
摄影：丁涛 Photo by：Ding Tao

叶 Leaves
摄影：丁涛 Photo by：Ding Tao

• 1~5 cm DBH + 5~20 cm DBH ○ ≥20 cm DBH
个体分布图 Distribution of individuals

径级分布表 DBH class

胸径区间 (Diameter class) (cm)	个体数 (No. of individuals in the plot)	比例 (Proportion) (%)
1~2	3	42.86
2~5	3	42.86
5~10	1	14.29
10~20	0	0.00
20~35	0	0.00
35~60	0	0.00
≥60	0	0.00

95 柔毛糙叶树

róu máo cāo yè shù | Pubescent Aphananthe

Aphananthe aspera (Thunb.) Planch. var. *pubescens* C. J. Chen
榆科 Ulmaceae

代码（SpCode）= APHASP
个体数（Individual number/15 hm^2）= 13
最大胸径（Max DBH）= 18.0 cm
重要值排序（Importance value rank）= 152

落叶乔木，高达 25 m。树皮带褐色或灰褐色，幼枝被伸展的灰色柔毛。叶纸质，卵形或卵状椭圆形，侧脉 6~10 对，伸达齿尖，叶背密被直立的柔毛，叶柄被伸展的灰色柔毛。核果近球形、椭圆形或卵状球形。花期 3~5 月，果期 8~10 月。

Deciduous trees, up to 25 m tall. Bark brown or grayish brown, branchlets with patent grayish pubescent. Leaf blade papery, ovate or ovate-elliptic, lateral veins 6–10 pairs, extending to margin, each ending in a tooth, abaxially with erect pubescent, petiole with patent grayish pubescent. Nuts subglobose, elliptic or obvate-globose. Fl. Mar.–May, fr. Aug.–Oct..

果枝 Fruiting branch
摄影：黄俞淞 Photo by：Huang Yusong

枝叶 Branch and leaves
摄影：黄俞淞 Photo by：Huang Yusong

果 Fruits
摄影：黄俞淞 Photo by：Huang Yusong

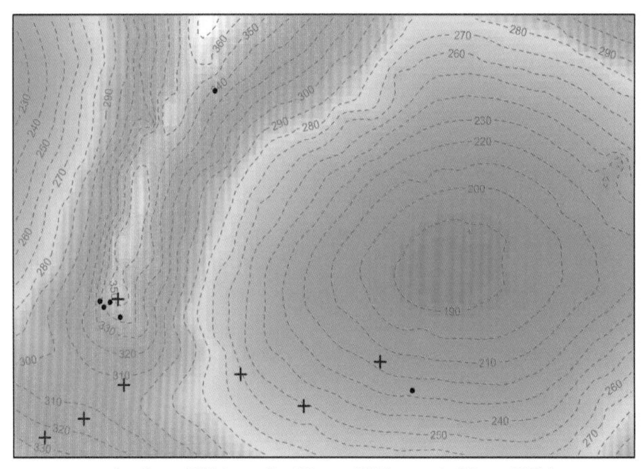

• 1~5 cm DBH + 5~20 cm DBH ○ ≥20 cm DBH
个体分布图 Distribution of individuals

径级分布表 DBH class

胸径区间 (Diameter class) (cm)	个体数 (No. of individuals in the plot)	比例 (Proportion) (%)
1~2	4	30.77
2~5	2	15.38
5~10	4	30.77
10~20	3	23.08
20~35	0	0.00
35~60	0	0.00
≥60	0	0.00

96 紫弹树 zǐ dàn shù | Biond's Hackberry

Celtis biondii Pamp.
榆科 Ulmaceae

代码（SpCode）= CELBIO
个体数（Individual number/15 hm^2）= 98
最大胸径（Max DBH）= 12.5 cm
重要值排序（Importance value rank）= 93

落叶小乔木至乔木，高可达 20 m。树皮暗灰色。叶宽卵形至卵状椭圆形，先端渐尖，基部偏斜，幼时两面被毛，老叶无毛，叶缘中上部有单锯齿或全缘。核果通常 2 个（稀 3 个）腋生，成熟时橙红色或带黑色。花期 4~5 月，果期 9~10 月。

Deciduous small trees to trees, up to 20 m tall. Bark gray. Leaf blade broadly ovate to ovate-elliptic, apex slenderly acuminate, base slightly oblique, both surfaces pubescent when young, glabrous when old, margin serrate on apical half, or entire. Drupe usually 2(3), axillary, orange red or blackish when mature. Fl. Apr.–May, fr. Sep.–Oct..

树干 Trunk
摄影：王斌 Photo by：Wang Bin

枝叶 Branch and leaves
摄影：黄俞淞 Photo by：Huang Yusong

果枝 Fruiting branch
摄影：许为斌 Photo by：Xu Weibin

• 1~5 cm DBH + 5~20 cm DBH ○ ≥20 cm DBH
个体分布图 Distribution of individuals

径级分布表 DBH class

胸径区间 (Diameter class) (cm)	个体数 (No. of individuals in the plot)	比例 (Proportion) (%)
1~2	24	24.49
2~5	33	33.67
5~10	35	35.71
10~20	6	6.12
20~35	0	0.00
35~60	0	0.00
≥60	0	0.00

97 朴树

pǔ shù | Chinese Hackberry

Celtis sinensis Pers.
榆科 Ulmaceae

代码（SpCode）＝ CELSIN
个体数（Individual number/15 hm^2）＝ 9
最大胸径（Max DBH）＝ 8.6 cm
重要值排序（Importance value rank）＝ 165

落叶乔木, 高 20 m。树皮灰色。叶卵形或卵状椭圆形，厚纸质，基部对称或稍偏斜，先端尖至渐尖，边近全缘或上部具齿。核果球形，直径 5~7 mm。花期 3~4 月，果期 9~10 月。

Deciduous trees, to 20 m tall. Bark gray. Leaf blade ovate or ovate-elliptic, thickly papery, base symmetric or moderately oblique, apex acute to shortly acuminate, margin subentire or crenate on apical half. Drupe globose, 5–7 mm in diam.. Fl. Mar.–Apr., fr. Sep.–Oct..

树干 Trunk
摄影：王斌 Photo by：Wang Bin

枝叶 Branch and leaves
摄影：王斌 Photo by：Wang Bin

叶 Leaf
摄影：王斌 Photo by：Wang Bin

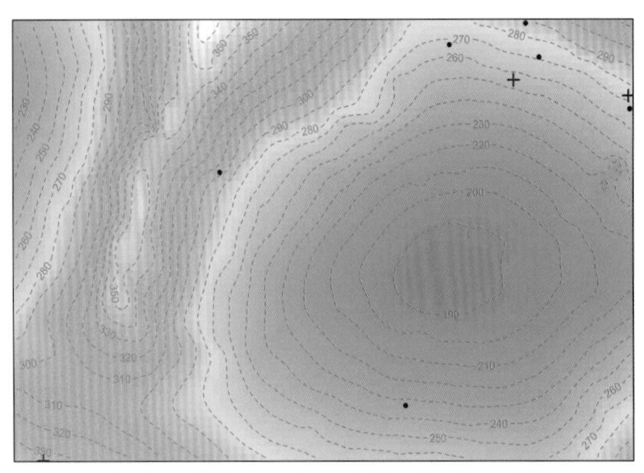

• 1~5 cm DBH + 5~20 cm DBH ○ ⩾20 cm DBH
个体分布图 Distribution of individuals

径级分布表 DBH class

胸径区间 (Diameter class) (cm)	个体数 (No. of individuals in the plot)	比例 (Proportion) (%)
1~2	2	22.22
2~5	4	44.44
5~10	3	33.33
10~20	0	0.00
20~35	0	0.00
35~60	0	0.00
⩾60	0	0.00

98 假玉桂

jiǎ yù guì | Timor Hackberry

Celtis timorensis Span.
榆科 Ulmaceae

代码（SpCode）= CELTIM
个体数（Individual number/15 hm^2）= 1122
最大胸径（Max DBH）= 38.4 cm
重要值排序（Importance value rank）= 14

常绿乔木，高 7~15 m。树皮灰白、灰色或灰褐色，木材有恶臭。叶卵状椭圆形或卵状长圆形，基部偏斜，侧脉每边 1~2 条。花淡紫色，组成具有 2~3 分枝的聚伞花序。核果宽卵形，成熟后红色至橙红色。花期 4~5 月，果期 10~11 月。

Evergreen trees, 7–15 m tall. Bark grayish white, gray, or grayish brown, timber rotten. Leaf blade ovate-elliptic or ovate-oblong, base oblique, lateral veins 1or 2 on each side of midvein. Flowers reddish, cyme 2-(3) branched. Drupe broadly ovoid, red to orange-red when mature. Fl. Apr.–May, fr. Oct.–Nov..

树干 Trunk
摄影：丁涛 Photo by：Ding Tao

嫩叶 New Leaves
摄影：王斌 Photo by：Wang Bin

果枝 Fruiting branch
摄影：黄俞淞 Photo by：Huang Yusong

• 1~5 cm DBH + 5~20 cm DBH ○ ≥20 cm DBH
个体分布图 Distribution of individuals

径级分布表 DBH class

胸径区间 (Diameter class) (cm)	个体数 (No. of individuals in the plot)	比例 (Proportion) (%)
1~2	276	24.60
2~5	398	35.47
5~10	284	25.31
10~20	138	12.30
20~35	25	2.23
35~60	1	0.09
≥60	0	0.00

99 常绿榆

chǎng lǜ yú | Lancet-leaf elm

Ulmus lanceifolia Roxb.
榆科 Ulmaceae

代码（SpCode）= ULMLAN
个体数（Individual number/15 hm^2）= 26
最大胸径（Max DBH）= 11.6 cm
重要值排序（Importance value rank）= 149

常绿小乔木。树皮灰褐色。叶边缘具单锯齿，叶面有光泽，叶背无毛；叶柄长 2~6 mm，近上面中脉有毛。花 3~7 朵簇生或排成簇状聚伞花序；翅果近圆形，无毛。花期和果期冬季或早春。

Evergreen small trees. Bark grayish brown. Margin simply serrate, abaxially glabrous, petiole 2–6 mm long, adaxially lustrous green and pubescent only on midvein. Inflorescences fascicled cymes, 3–7-flowered. Samaras ovbicular, glabrous. Fl. and fr. winter or early spring.

树干 Trunk
摄影：王斌 Photo by: Wang Bin

叶 Leaves
摄影：黄俞淞 Photo by: Huang Yusong

果序 Infructescence
摄影：黄俞淞 Photo by: Huang Yusong

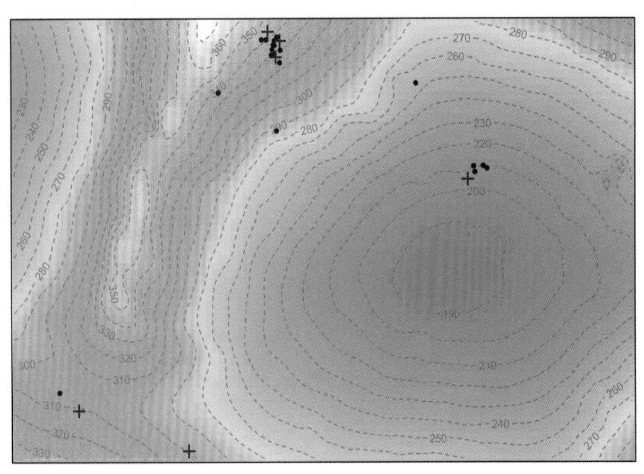

• 1~5 cm DBH + 5~20 cm DBH ○ ≥20 cm DBH
个体分布图 Distribution of individuals

径级分布表 DBH class

胸径区间 (Diameter class) (cm)	个体数 (No. of individuals in the plot)	比例 (Proportion) (%)
1~2	8	30.77
2~5	12	46.15
5~10	5	19.23
10~20	1	3.85
20~35	0	0.00
35~60	0	0.00
≥60	0	0.00

100 白桂木

bái guì mù | Silver-back Artocarpus

Artocarpus hypargyreus Hance
桑科 Moraceae

代码（SpCode） = ARTHYP
个体数（Individual number/15 hm^2） = 2
最大胸径（Max DBH） = 13.4 cm
重要值排序（Importance value rank） = 201

乔木，高 10~25 m。树皮深紫色，片状剥落。叶革质，椭圆形至倒卵形，全缘，幼树之叶常为羽状浅裂，网脉很明显，干时背面灰白色；叶柄被毛。花序单生叶腋。聚花果近球形，直径 3~4 cm，表面被褐色柔毛，微具乳头状凸起。花期春夏。

Trees, 10–25 m tall. Bark dark purple, exfoliating. Leaf blade leathery, elliptic to obovate, entire, leaves of young trees usually pinnatilobate, veins conspicuous, grayish white when dry, petiole pubescent. Inflrescences axillary, solitary. Fruiting syncarp, 3–4 cm in diam., brownish pubescent, papillate. Fl. spring to summer.

枝叶 Branch and leaves
摄影：黄俞淞 Photo by：Huang Yusong

果 Fruit
摄影：黄俞淞 Photo by：Huang Yusong

叶背 Leaf back
摄影：黄俞淞 Photo by：Huang Yusong

• 1~5 cm DBH + 5~20 cm DBH ○ ≥20 cm DBH
个体分布图 Distribution of individuals

径级分布表 DBH class

胸径区间 (Diameter class) (cm)	个体数 (No. of individuals in the plot)	比例 (Proportion) (%)
1~2	0	0.00
2~5	1	50.00
5~10	0	0.00
10~20	1	50.00
20~35	0	0.00
35~60	0	0.00
≥60	0	0.00

101 胭脂 yān zhī | Tonkin Artocarpus

Artocarpus tonkinensis A. Chev. ex Gagnep.
桑科 Moraceae

代码（SpCode）= ARTTON
个体数（Individual number/15 hm^2）= 29
最大胸径（Max DBH）= 32.1 cm
重要值排序（Importance value rank）= 120

乔木，高 14~16 m。小枝淡红褐色，常被平伏短柔毛。叶革质，椭圆形或倒卵形，全缘，先端有浅锯齿，背面密被微柔毛。花序单生叶腋，雄花序倒卵圆形，雌花序球形。聚花果近球形，直径达 6.5 cm，成熟时黄色，干后红褐色。花期夏秋，果秋冬季。

Trees, 14–16 m tall. Branchlets pale reddish brown, curly to appressed puberulent. Leaf blade leathery or obovate, margin entire, apically with a few shallow teeth, abaxially densely pubescent. Inflorescences axially, solitary, male inflorecences oblong, female infrorcccnccs globous. Fruiting syncarp, globous, ca. 6.5 cm in diam. yellow when mature, brownish red when dry. Fl. summer to autumn, fr. winter.

枝叶 Branch and leaves
摄影：黄俞淞 Photo by: Huang Yusong

叶 Leaf
摄影：黄俞淞 Photo by: Huang Yusong

嫩叶 New Leaf
摄影：黄俞淞 Photo by: Huang Yusong

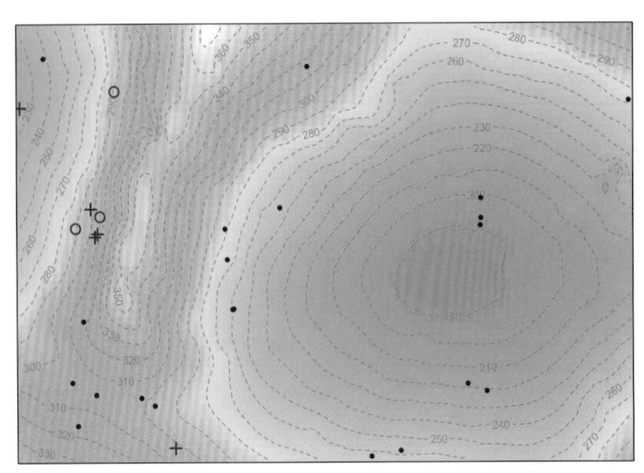

● 1~5 cm DBH + 5~20 cm DBH ○ ≥20 cm DBH
个体分布图 Distribution of individuals

径级分布表 DBH class

胸径区间 (Diameter class) (cm)	个体数 (No. of individuals in the plot)	比例 (Proportion) (%)
1~2	9	31.03
2~5	12	41.38
5~10	2	6.90
10~20	3	10.34
20~35	3	10.34
35~60	0	0.00
≥60	0	0.00

102 构树

gòu shù | Paper Mulberry

Broussonetia papyrifera (Linn.) L'Hér. ex Vent.
桑科 Moraceae

代码（SpCode）= BROPAP
个体数（Individual number/15 hm^2）= 4
最大胸径（Max DBH）= 26.6 cm
重要值排序（Importance value rank）= 167

乔木，高 10~20 m。树皮暗灰色，小枝密生柔毛。叶广卵形至长椭圆状卵形，基部心形，两侧稍不对称，边缘具粗锯齿，不分裂或 3~5 裂。花雌雄异株，雄花序为柔荑花序，雌花序球形头状。聚花果直径 1.5~3 cm，成熟时橙红色，肉质。花期 4~5 月，果期 6~7 月。

Trees, 10–20 m tall. Bark dark gray, branchlets densely pubescent. Leaf blade broadly ovate to narrowly elliptic-ovate, base cordate, asymmetric, margin coarsely serrate, simple or 3–5-lobed. Flowers dioecism, male inflorescences ament, female inflorescences globose. Collective fruit 1.5–3 cm in diam., orange red when mature, fleshy. Fl. Apr.–May, fr. Jun.–Jul..

树干 Trunk
摄影：黄俞淞 Photo by：Huang Yusong

果枝 Fruiting branches
摄影：黄俞淞 Photo by：Huang Yusong

聚花果 Collective fruits
摄影：黄俞淞 Photo by：Huang Yusong

• 1~5 cm DBH　+ 5~20 cm DBH　○ ≥20 cm DBH
个体分布图 Distribution of individuals

径级分布表 DBH class

胸径区间 (Diameter class) (cm)	个体数 (No. of individuals in the plot)	比例 (Proportion) (%)
1~2	0	0.00
2~5	0	0.00
5~10	0	0.00
10~20	2	50.00
20~35	2	50.00
35~60	0	0.00
≥60	0	0.00

103 大果榕

dà guǒ róng | Roxburgh Fig

Ficus auriculata Lour.

桑科 Moraceae

代码（SpCode）= FICAUR

个体数（Individual number/15 hm^2）= 24

最大胸径（Max DBH）= 29.3 cm

重要值排序（Importance value rank）= 119

落叶乔木或小乔木，高 4~10 m。树皮灰褐色，粗糙。叶互生，厚纸质，广卵状心形；雄花着生在小梗上或无柄，花被片 3，雌花花被片 3 裂；榕果簇生于树干基部或老茎短枝上，梨形，具明显的纵棱 8~12 条。花期 8 月至翌年 3 月，果期 5~8 月。

Deciduous trees or small trees, 4–10 m tall. Bark grayish brown, rough. Leaf blade alternate, thick papery, broadly ovate-cordate. Male flowers pedicellate or sessile, calyx 3, female flowers 3-lobed. Figs on specialized leafless branchlets at base of trunk and main branches, pear-shaped, with 8–12 conspicuous longitudinal ridges. Fl. Aug.–Mar., fr. May–Aug. of next year.

果 Fruits

摄影：刘晟源 Photo by：Liu Shengyuan

枝叶 Branch and leaves

摄影：黄俞淞 Photo by：Huang Yusong

嫩叶 New Leaves

摄影：黄俞淞 Photo by：Huang Yusong

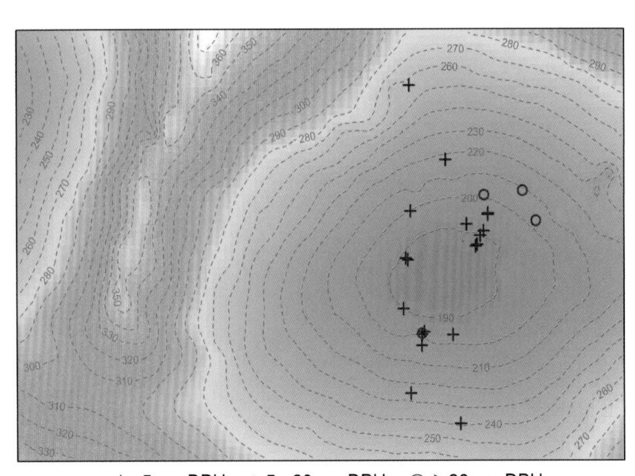

• 1~5 cm DBH + 5~20 cm DBH ○ ≥20 cm DBH

个体分布图 Distribution of individuals

径级分布表 DBH class

胸径区间 (Diameter class) (cm)	个体数 (No. of individuals in the plot)	比例 (Proportion) (%)
1~2	0	0.00
2~5	0	0.00
5~10	6	25.00
10~20	14	58.33
20~35	4	16.67
35~60	0	0.00
≥60	0	0.00

104 楔叶榕

xiē yè róng | Wedge-leaf Fig

Ficus trivia Corner
桑科 Moraceae

代码（SpCode）= FICTRI
个体数（Individual number/15 hm^2）= 18
最大胸径（Max DBH）= 8.5 cm
重要值排序（Importance value rank）= 150

灌木或小乔木，高 3~8 m。小枝红褐色。叶纸质，卵状椭圆形至倒卵形，基部钝圆或宽楔形，全缘，背面基部脉腋具腺体，叶柄长 2~5 cm，托叶卵状披针形。榕果成对腋生或单生，成熟时红色至紫色，近球形，无毛。瘦果卵球形，光滑。花期 9 月至翌年 4 月，果期 5~8 月。

Shrubs or small trees, 3–8 m tall. Branchlets reddish brown. Leaf blade papery, ovate-elliptic to obovate, base broadly cuneate to rounded, margin entire, axils of basal secondary veins abaxially glandular, petiole 2–5 cm long, stipules ovate-lanceolate. Figs axillary, solitary or paired, red to purple when mature, subglobose, glabrous. Achenes ovoid, smooth. Fl. Sep.–Apr. of next year, fr. May–Aug..

植株 Whole plant
摄影：黄俞淞 Photo by：Huang Yusong

叶 Leaves
摄影：黄俞淞 Photo by：Huang Yusong

果 Fruits
摄影：黄俞淞 Photo by：Huang Yusong

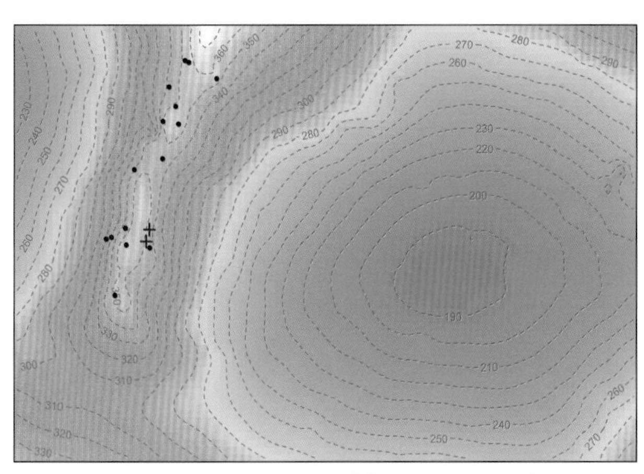

• 1~5 cm DBH　+ 5~20 cm DBH　○ ≥20 cm DBH
个体分布图 Distribution of individuals

径级分布表 DBH class

胸径区间 (Diameter class) (cm)	个体数 (No. of individuals in the plot)	比例 (Proportion) (%)
1~2	8	44.44
2~5	8	44.44
5~10	2	11.11
10~20	0	0.00
20~35	0	0.00
35~60	0	0.00
≥60	0	0.00

105 歪叶榕 wāi yè róng | Oblique-leaf Fig

Ficus cyrtophylla (Wall. ex Miq.) Miq.
桑科 Moraceae

代码（SpCode）= FICCYR
个体数（Individual number/15 hm^2）= 14
最大胸径（Max DBH）= 23.1 cm
重要值排序（Importance value rank）= 144

常绿灌木或小乔木，高 3~6 m。树皮灰色，光滑。叶二列，互生，两侧极不对称，表面极粗糙，具乳突状钟乳体。雄花花被片 4，雌花花被片 5。榕果成对或簇生叶腋。花期 5~6 月。

Evergreen shrubs or small trees, 3–6 m tall. Bark gray, smooth. Leaves distichous, alternate, both sides strongly asymmetric, adaxially very rough, with papillate cystoliths. Female flowers: calyx lobes 4, male flowers: calyx lobes 5. Figs pairs or fascicled, axillary. Fl. May–Jun..

果枝 Fruiting branches
摄影：黄俞淞 Photo by: Huang Yusong

果 Fruits
摄影：丁涛 Photo by: Ding Tao

叶 Leaves
摄影：黄俞淞 Photo by: Huang Yusong

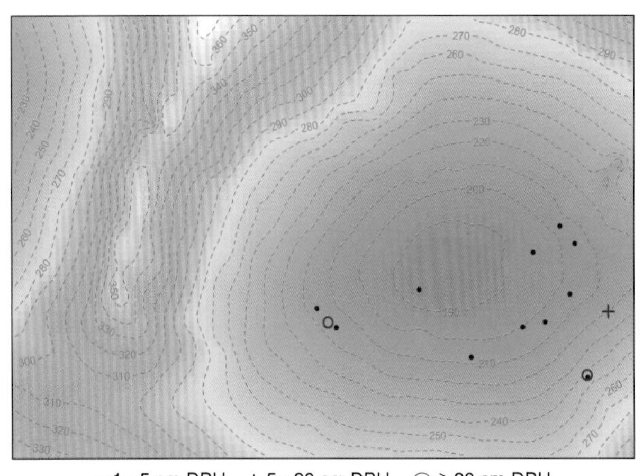
• 1~5 cm DBH + 5~20 cm DBH ○ ≥20 cm DBH
个体分布图 Distribution of individuals

径级分布表 DBH class

胸径区间 (Diameter class) (cm)	个体数 (No. of individuals in the plot)	比例 (Proportion) (%)
1~2	5	35.71
2~5	6	42.86
5~10	1	7.14
10~20	0	0.00
20~35	2	14.29
35~60	0	0.00
≥60	0	0.00

106 矮小天仙果

ǎi xiǎo tiān xiān guǒ | Erect Fig

Ficus erecta Thunb.
桑科 Moraceae

代码（SpCode）= FICERE
个体数（Individual number/15 hm^2）= 2
最大胸径（Max DBH）= 1.8 cm
重要值排序（Importance value rank）= 191

落叶或半落叶灌木，高 3~4 m。树皮棕灰色。叶倒卵形至狭倒卵形，先端急尖，具短尖头，基部圆形或浅心形，表面无毛，微粗糙, 叶柄长 1.5~4 cm。榕果单生叶腋，球形，无毛，直径 1~1.5 cm，成熟时红色。花果期 5 月至 6 月。

Deciduous or semideciduous shrubs, 3–4 m tall. Bark grayish brown. Leaf blade obovate to narrowly obovate, apex shortly acuminate or acute and mucronate, base rounded or cordate, glabrous adaxially, sparsely rough, petiole 1.5–4 cm long. Figs axillary, solitary, globose, glabrous, 1–1.5 cm in diam., red when mature. Fl. and fr. May–Jun.

果枝 Fruiting branch
摄影：黄俞淞 Photo by：Huang Yusong

果 Fruit
摄影：黄俞淞 Photo by：Huang Yusong

叶 Leaves
摄影：黄俞淞 Photo by：Huang Yusong

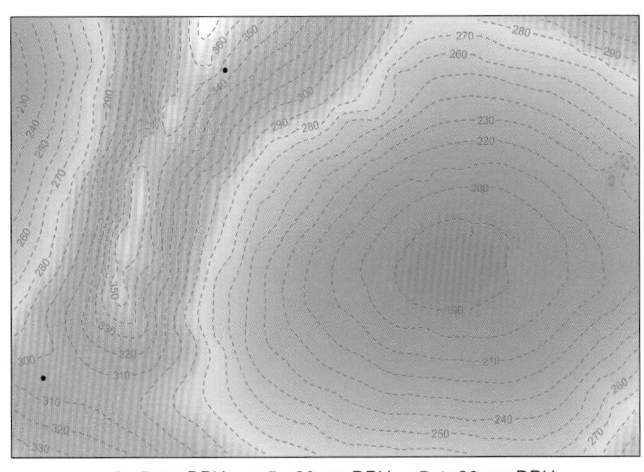

• 1~5 cm DBH + 5~20 cm DBH ○ ≥20 cm DBH
个体分布图 Distribution of individuals

径级分布表 DBH class

胸径区间 (Diameter class) (cm)	个体数 (No. of individuals in the plot)	比例 (Proportion) (%)
1~2	2	100.00
2~5	0	0.00
5~10	0	0.00
10~20	0	0.00
20~35	0	0.00
35~60	0	0.00
≥60	0	0.00

107 大叶水榕

dà yè shuǐ róng | Vary-Glabrous Fig

Ficus glaberrima Blume
桑科 Moraceae

代码（SpCode）= FICGLA
个体数（Individual number/15 hm²）= 99
最大胸径（Max DBH）= 48.2 cm
重要值排序（Importance value rank）= 61

常绿乔木，高约 15 m。树皮灰色，小枝幼时微被柔毛。叶薄革质，全缘，基出侧脉短。雄花、雌花和瘿花同生于一榕果内，雄花花被片 4，瘿花花被 4 深裂，雌花花被片 4。榕果成对腋生，球形，成熟时橙黄色，花果期 5~9 月。

Evergreen trees, to 15 m tall. Bark gray. Branchlets pubescent when young. Leaves thinly leathery, margin entire, basal lateral veins short. Mall, gall, and female flowers within same fig, male flowers: calyx lobes 4, gall flowers: calyx deeply 4-lobed, female flowers: calyx 4. Figs paired, axillary, globose, orange-yellow when maturity. Fl. and fr. May–Sep..

枝叶 Branch and leaves
摄影：黄俞淞 Photo by: Huang Yusong

果序 Infructescence
摄影：黄俞淞 Photo by: Huang Yusong

叶 Leaf
摄影：黄俞淞 Photo by: Huang Yusong

• 1~5 cm DBH + 5~20 cm DBH ○ ≥20 cm DBH
个体分布图 Distribution of individuals

径级分布表 DBH class

胸径区间 (Diameter class) (cm)	个体数 (No. of individuals in the plot)	比例 (Proportion) (%)
1~2	18	18.18
2~5	36	36.36
5~10	15	15.15
10~20	20	20.20
20~35	9	9.09
35~60	1	1.01
≥60	0	0.00

108 对叶榕

duì yè róng | Opposite-leaf Fig

Ficus hispida L. f.
桑科 Moraceae

代码（SpCode）= FICHIS
个体数（Individual number/15 hm^2）= 2989
最大胸径（Max DBH）= 31.5 cm
重要值排序（Importance value rank）= 11

落叶灌木或小乔木。通常茎中空。叶通常对生，厚纸质，卵状长椭圆形或倒卵状矩圆形，表面粗糙，托叶 2。雌花和瘿花均无花被。榕果腋生或生于落叶枝上，陀螺形，成熟后黄色。花果期 6~7 月。

Deciduous shrubs or samll trees. Stem usually hollow. Leaf blade usually opposite, thickly papery, ovate-oblong or obovate-oblong, adaxially rough, stipule 2. Calyx of female flowers and gall flowers absent. Figs axillary or on leafless branchlets, top-shaped, yellow when mature. Fl. and fr. Jun.–Jul..

树干 Trunk
摄影：黄俞淞 Photo by：Huang Yusong

枝叶 Branch and leaves
摄影：黄俞淞 Photo by：Huang Yusong

果序 Infructescence
摄影：黄俞淞 Photo by：Huang Yusong

● 1~5 cm DBH + 5~20 cm DBH ○ ≥20 cm DBH
个体分布图 Distribution of individuals

径级分布表 DBH class

胸径区间 (Diameter class) (cm)	个体数 (No. of individuals in the plot)	比例 (Proportion) (%)
1~2	1571	52.56
2~5	899	30.08
5~10	386	12.91
10~20	130	4.35
20~35	3	0.10
35~60	0	0.00
≥60	0	0.00

109 榕树

róng shù | Small-fruit Fig

Ficus microcarpa L. f.
桑科 Moraceae

代码（SpCode）= FICMIC
个体数（Individual number/15 hm^2）= 72
最大胸径（Max DBH）= 65.0 cm
重要值排序（Importance value rank）= 60

常绿乔木，高可达 15~30 m。具气生根，树皮深灰色。叶薄革质，倒卵形至卵形，全缘。榕果成对腋生，近球形，成熟时黄或微红色。花期 5~6 月。

Evergreen trees, up to 15–30 m. Roots aerial, bark gray. Leaves thinly leathery, obovate to ovate, margins entire. Figs paired, axillary, nearly globose, yellow or thinly red when maturity. Fl. May–Jun..

树干 Trunk
摄影：王斌 Photo by：Wang Bin

果 Fruits
摄影：林春蕊 Photo by：Lin Chunrui

叶 Leaves
摄影：林春蕊 Photo by：Lin Chunrui

• 1~5 cm DBH + 5~20 cm DBH ○ ≥20 cm DBH
个体分布图 Distribution of individuals

径级分布表 DBH class

胸径区间 (Diameter class) (cm)	个体数 (No. of individuals in the plot)	比例 (Proportion) (%)
1~2	15	20.83
2~5	12	16.67
5~10	11	15.28
10~20	17	23.61
20~35	13	18.06
35~60	3	4.17
≥60	1	1.39

110 苹果榕

píng guǒ róng | Few-Teeth Fig

Ficus oligodon Miq.
桑科 Moraceae

代码（SpCode）= FICOLI
个体数（Individual number/15 hm²）= 297
最大胸径（Max DBH）= 28.6 cm
重要值排序（Importance value rank）= 49

乔木，高 5~10 m。树皮灰色。叶互生，纸质，倒卵椭圆形或椭圆形，边缘在叶片 2/3 以上具不规则粗锯齿，表面无毛，背面密生小瘤体。榕果簇生于老茎发出的短枝上，梨形或近球形，表面有 4~6 条纵棱和小瘤体，被微柔毛，成熟深红色。花期 9 月至翌年 4 月，果期 5~6 月。

Trees, 5–10 m tall. Bark gray. Leaves alternate, papery, obovate-elliptic ot elliptic, margin Irregularly toothed on apical 2/3, adaxially glabrous, abaxially densely small tuberculate. Figs clustered on short branchlets of old stems, pear-shaped or globose, with 4–6 longitudinalridges and small tubercles, pubescent, dark red when mature. Fl. Sep.–Apr. of next year, fr. May–Jun..

树干 Trunk
摄影：黄俞淞 Photo by：Huang Yusong

果序 Infructescence
摄影：黄俞淞 Photo by：Huang Yusong

叶 Leaves
摄影：黄俞淞 Photo by：Huang Yusong

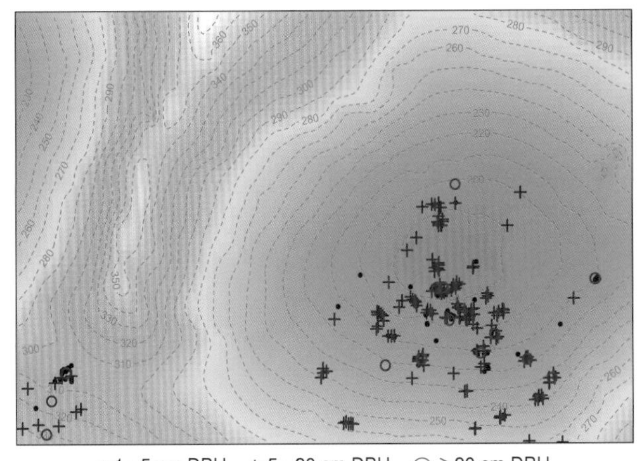

● 1~5 cm DBH ＋ 5~20 cm DBH ○ ≥20 cm DBH
个体分布图 Distribution of individuals

径级分布表 DBH class

胸径区间 (Diameter class) (cm)	个体数 (No. of individuals in the plot)	比例 (Proportion) (%)
1~2	43	14.48
2~5	76	25.59
5~10	61	20.54
10~20	103	34.68
20~35	14	4.71
35~60	0	0.00
≥60	0	0.00

111 直脉榕

zhí mài róng | Straight-Nerve Fig

Ficus orthoneura Lévl. et Vant.
桑科 Moraceae

代码（SpCode）= FICORT
个体数（Individual number/15 hm^2）= 19
最大胸径（Max DBH）= 18.6 cm
重要值排序（Importance value rank）= 132

小乔木，高 2~10 m。小枝圆柱形。叶生小枝顶端，革质，全缘，倒卵圆形或椭圆形，侧脉 7~15 对，平行直出，至边缘弯拱向上网结。榕果成对或单生叶腋，球形或倒卵状球形。花期 4~9 月。

Small trees, 2–10 m tall. Branchlets terete. Leaves clustered apically on branchlets, leathery, obovate or elliptic, lateral veins 7–15 pairs, parallel, straight,and reticulate near margin. Figs axillary on leafy branchlets, paired or solitary, globose or obovate-globose. Fl. Apr.–Sep..

果枝 Fruiting branch
摄影：黄俞淞 Photo by：Huang Yusong

果序 Infructescence
摄影：黄俞淞 Photo by：Huang Yusong

叶 Leaves
摄影：黄俞淞 Photo by：Huang Yusong

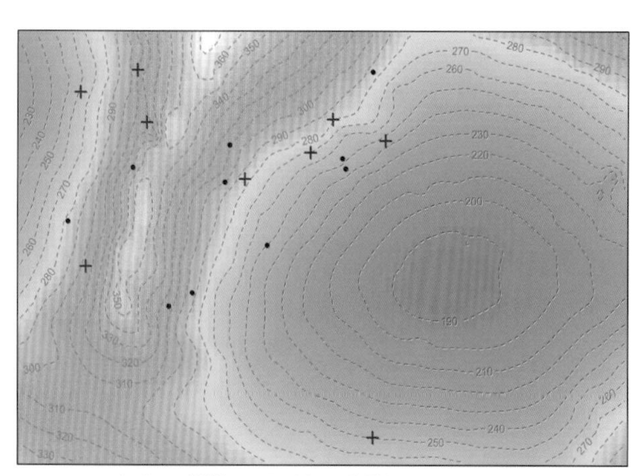

• 1~5 cm DBH　+ 5~20 cm DBH　○ ≥20 cm DBH
个体分布图 Distribution of individuals

径级分布表 DBH class

胸径区间 (Diameter class) (cm)	个体数 (No. of individuals in the plot)	比例 (Proportion) (%)
1~2	2	10.53
2~5	8	42.11
5~10	7	36.84
10~20	2	10.53
20~35	0	0.00
35~60	0	0.00
≥60	0	0.00

112 斜叶榕

xié yè róng | Gibbous Dye Fig

Ficus tinctoria G. Forst. subsp. *gibbosa* (Blume) Corner
桑科 Moraceae

代码（SpCode）= FICTIN
个体数（Individual number/15 hm^2）= 164
最大胸径（Max DBH）= 23.6 cm
重要值排序（Importance value rank）= 56

常绿小乔木。树皮微粗糙，小枝褐色。叶薄革质，椭圆形、卵状椭圆形至菱形，基部宽楔形，偏斜。雄花花被片 4-6，瘿花与雄花花被相似，雌花花被片 4，榕果球形，单生或成对腋生。花果期 6~7 月。

Evergreen small trees. Bark slightly rough, branchlets brown. Leaf brade slightly leathery, elliptic、ovate-elliptic to rhombic, base broadly cuneate, asymmetric. Male flowers: calyx lobes 4–6, similar to gall flowers. Female flowers: calyx lobes 4. Figs globose, solitary or pair axillary. Fl. and fr. Jun.–Jul..

果枝 Fruiting branch
摄影：黄俞淞 Photo by：Huang Yusong

叶 Leaves
摄影：黄俞淞 Photo by：Huang Yusong

果 Fruits
摄影：黄俞淞 Photo by：Huang Yusong

• 1~5 cm DBH + 5~20 cm DBH ○ ≥20 cm DBH
个体分布图 Distribution of individuals

径级分布表 DBH class

胸径区间 (Diameter class) (cm)	个体数 (No. of individuals in the plot)	比例 (Proportion) (%)
1~2	37	22.56
2~5	46	28.05
5~10	45	27.44
10~20	33	20.12
20~35	3	1.83
35~60	0	0.00
≥60	0	0.00

113 光叶榕

guāng yè róng | Smooth-leaf Fig

Ficus laevis Blume
桑科 Moraceae

代码（SpCode）= FICLAE
个体数（Individual number/15 hm^2）= 1
最大胸径（Max DBH）= 12.8 cm
重要值排序（Importance value rank）= 203

攀援藤状灌木，通常光滑无毛。叶螺旋状排列，膜质，圆形至宽卵形，全缘，先端钝或具短尖，基部圆形至浅心形，叶柄长 3.5-7 cm。榕果单生或成对腋生，球形，幼时绿色，成熟紫色。花果期 4-6 月。

Shrubs, scandent, usually glabrous. Leaves spirally arranged, membranous, leaf blade rounded to broadly ovate, margin entire, apex obtuse or mucronate, base rounded to slightly cordate, petiole 3.5–7 cm long. Figs axillary, solitary or paired, globous, green when young, purple when mature. Fl. and fr. Apr. Jun..

果枝 Fruiting branch
摄影：黄俞淞 Photo by：Huang Yusong

果 Fruit
摄影：黄俞淞 Photo by：Huang Yusong

叶 Leaf
摄影：黄俞淞 Photo by：Huang Yusong

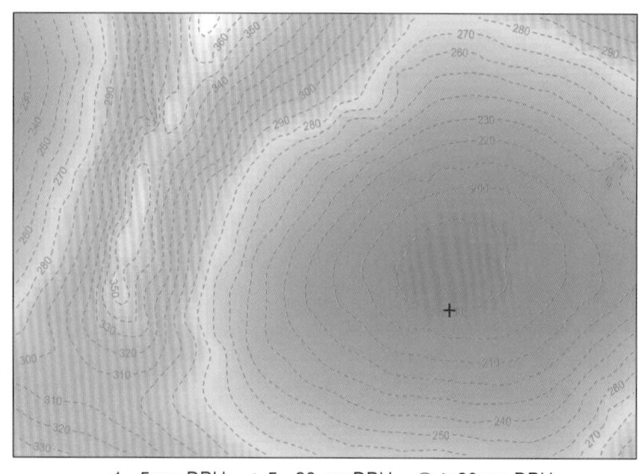

• 1～5 cm DBH + 5～20 cm DBH ○ ≥20 cm DBH
个体分布图 Distribution of individuals

径级分布表 DBH class

胸径区间 (Diameter class) (cm)	个体数 (No. of individuals in the plot)	比例 (Proportion) (%)
1～2	0	0.00
2～5	0	0.00
5～10	0	0.00
10～20	1	100.00
20～35	0	0.00
35～60	0	0.00
≥60	0	0.00

114 黄葛树

huáng gě shù | White Fig

Ficus virens Ait.
桑科 Moraceae

代码（SpCode）= FICVIR
个体数（Individual number/15 hm^2）= 88
最大胸径（Max DBH）= 35.1 cm
重要值排序（Importance value rank）= 71

落叶或半落叶乔木，有板根或支柱根。叶卵状披针形至椭圆状卵形，全缘，基生叶脉短，侧脉 7~10 对，背面突起。榕果单生或成对腋生或簇生于已落叶枝叶腋，球形，成熟时紫红色。花期 5~8 月。

Deciduous or semideciduous trees, with buttress or prop roots. Leaf blade ovate-lanceolate to elliptic-ovate, entire, basal lateral veins short, secondary veins 7–10 pairs, abaxially prominent. Figs solitary or pairs, or in clusters on leafless older branchlets, globose, purple red when mature. Fl. May–Aug..

嫩枝 New branch
摄影：黄俞淞 Photo by：Huang Yusong

果 Fruit
摄影：黄俞淞 Photo by：Huang Yusong

叶 Leaves
摄影：黄俞淞 Photo by：Huang Yusong

• 1~5 cm DBH + 5~20 cm DBH ○ ≥20 cm DBH
个体分布图 Distribution of individuals

径级分布表 DBH class

胸径区间 (Diameter class) (cm)	个体数 (No. of individuals in the plot)	比例 (Proportion) (%)
1~2	15	17.05
2~5	25	28.41
5~10	21	23.86
10~20	20	22.73
20~35	6	6.82
35~60	1	1.14
≥60	0	0.00

115 柘

zhè | Cudrang

Maclura tricuspidata Carrière
桑科 Moraceae

代码（SpCode）= MACTRI
个体数（Individual number/15 hm^2）= 20
最大胸径（Max DBH）= 9.2 cm
重要值排序（Importance value rank）= 145

落叶灌木或小乔木，高 1~7 m。树皮灰褐色，小枝有棘刺。叶卵形或菱状卵形，偶为三裂。雌雄异株，雌雄花序均为球形头状花序，单生或成对腋生。聚花果近球形，肉质，成熟时橘红色。花期 5~6 月，果期 6~7 月。

Deciduous shrubs or small trees, 1–7 m tall. Bark grayish brown, branchlets ridged. Leaf blade ovate to rhombic-ovate, occasionally 3-lobed. Dioecism, male and female inflorescences capitulum, solitary or paired axillary. Figs subglobose,fleshy, yellow red when mature. Fl. May–Jun., fr. Jun.–Jul..

果枝 Fruiting branch
摄影：刘晟源 Photo by：Liu Shengyuan

枝叶 Branch and leaves
摄影：黄俞淞 Photo by：Huang Yusong

聚花果 Collective fruit
摄影：黄俞淞 Photo by：Huang Yusong

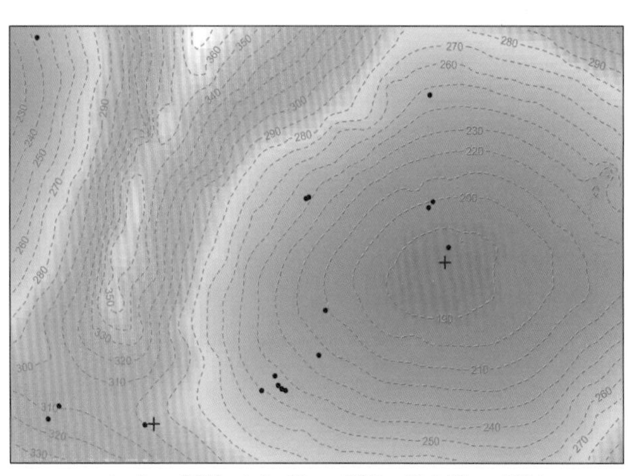

• 1~5 cm DBH + 5~20 cm DBH ○ ≥20 cm DBH
个体分布图 Distribution of individuals

径级分布表 DBH class

胸径区间 (Diameter class) (cm)	个体数 (No. of individuals in the plot)	比例 (Proportion) (%)
1~2	11	55.00
2~5	7	35.00
5~10	2	10.00
10~20	0	0.00
20~35	0	0.00
35~60	0	0.00
≥60	0	0.00

116 奶桑

nǎi sāng | Big-tail Mulberry

Morus macroura Miq.
桑科 Moraceae

代码（SpCode）= MORMAC
个体数（Individual number/15 hm^2）= 1
最大胸径（Max DBH）= 11.0 cm
重要值排序（Importance value rank）= 205

小乔木，高 7~12 m。小枝幼时被柔毛。叶膜质，卵形或宽卵形，边缘具细密锯齿，基生侧脉延长至叶片中部，侧脉 4~6 对。花雌雄异株，雄花序穗状，单生或成对腋生。聚花果成熟时黄白色。花期 3~4 月，果期 4~5 月。

Small trees, 7–12 m tall. Branchlets pubescent when young. Leaf blade menbranous, ovate or bloadly ovate, margin densely serrate, basal lateral veins along to middle, lateral veins 4–6 pairs. Dioecism, male inflorescences spicate, solitary or paired axillary. Figs yellowish white. Fl. Mar.–Apr., fr. Apr.–May.

树干 Trunk
摄影：黄俞淞 Photo by：Huang Yusong

叶 Leaf
摄影：黄俞淞 Photo by：Huang Yusong

枝叶 Branch and leaves
摄影：黄俞淞 Photo by：Huang Yusong

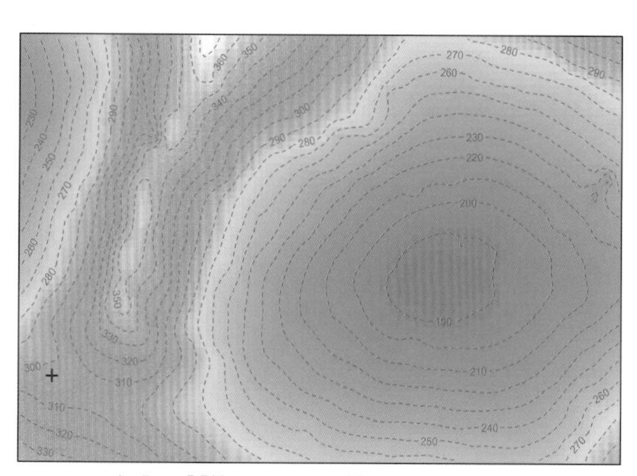

• 1~5 cm DBH + 5~20 cm DBH ○ ≥20 cm DBH
个体分布图 Distribution of individuals

径级分布表 DBH class

胸径区间 (Diameter class) (cm)	个体数 (No. of individuals in the plot)	比例 (Proportion) (%)
1~2	0	0.00
2~5	0	0.00
5~10	0	0.00
10~20	1	100.00
20~35	0	0.00
35~60	0	0.00
≥60	0	0.00

117 米扬噎

mǐ yáng yē | Tonkin Streblus

Streblus tonkinensis (Dubard et Eberh.) Corner

桑科 Moraceae

代码（SpCode）= STRTON

个体数（Individual number/15 hm^2）= 995

最大胸径（Max DBH）= 23.4 cm

重要值排序（Importance value rank）= 28

常绿灌木或小乔木，高 2~12 m。叶纸质，倒卵状长圆形，先端尾状渐尖，尾尖歪向一边，叶背具细小瘤点。花单性，雌雄同株或同序，雄花序近球形，腋生，雌花单生叶腋或生于雄花序中部。核果近球形。花期春夏。

Evergreen shrubs or small trees, 2–12 m tall. Leaf blade papery, obovate-oblong, apex caudate to acuminate and with an asymmetric acumen, abaxially densely covered with small tubercles. Flowers solitary, momoecism or androgynous, male inflorescences globose, axillary, female flowers axillary, solitary or at the middle of male inflorescences. Drupes subglobose. Fl. spring to summer.

树干 Trunk

摄影：王斌 Photo by: Wang Bin

花枝 Flowering branches

摄影：黄俞淞 Photo by: Huang Yusong

果枝 Fruiting branch

摄影：黄俞淞 Photo by: Huang Yusong

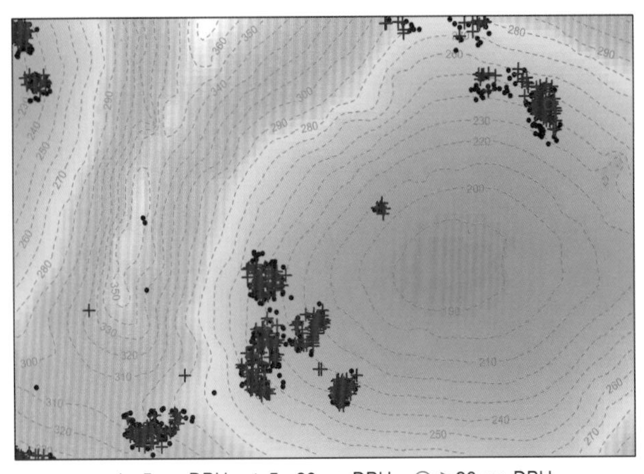

• 1~5 cm DBH + 5~20 cm DBH ○ ≥20 cm DBH

个体分布图 Distribution of individuals

径级分布表 DBH class

胸径区间 (Diameter class) (cm)	个体数 (No. of individuals in the plot)	比例 (Proportion) (%)
1~2	291	29.25
2~5	328	32.96
5~10	288	28.94
10~20	86	8.64
20~35	2	0.20
35~60	0	0.00
≥60	0	0.00

118 青叶苎麻

qīng yè zhù má | Green-leaf False-nettle

Boehmeria nivea var. *tenacissima* (Gaudich.) Miq.
荨麻科 Urticaceae

代码（SpCode）= BOENIV
个体数（Individual number/15 hm^2）= 619
最大胸径（Max DBH）= 10.6 cm
重要值排序（Importance value rank）= 36

亚灌木或灌木，高 0.5~1.5 m。茎和叶柄密或疏被短伏毛。叶片多为卵形或椭圆状卵形，顶端长渐尖，基部多为圆形，或为宽楔形，下面疏被短伏毛，绿色，或有薄层白色毡毛。圆锥花序腋生。花期 8~10 月。

Subshrubs or shrubs, 0.5–1.5 m tall. Stems and petioles densely or sparsely appressed strigose. Leaf blade usually ovate or elliptic-ovate, apex long caudate, base usually rounded, or broadly cuniform, abaxially sparsely appressed strigose, green or with white felt hair. Panicle axillary. Fl. Aug.–Oct..

花枝 Flowering branches
摄影：黄俞淞 Photo by：Huang Yusong

花序 Inflorescence
摄影：黄俞淞 Photo by：Huang Yusong

果序 Infructescence
摄影：黄俞淞 Photo by：Huang Yusong

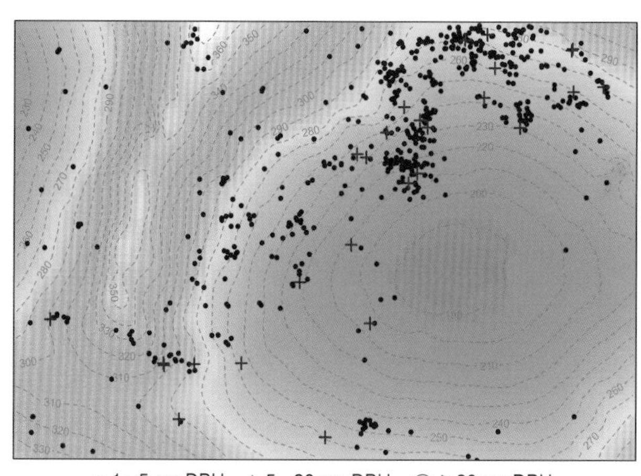

• 1~5 cm DBH + 5~20 cm DBH ○ ≥20 cm DBH
个体分布图 Distribution of individuals

径级分布表 DBH class

胸径区间 (Diameter class) (cm)	个体数 (No. of individuals in the plot)	比例 (Proportion) (%)
1~2	339	54.77
2~5	253	40.87
5~10	25	4.04
10~20	2	0.32
20~35	0	0.00
35~60	0	0.00
≥60	0	0.00

119 葡萄叶艾麻

pú táo yè ài má | Grape-leaf Woodnettle

Laportea violacea Gagnep.
荨麻科 Urticaceae

代码（SpCode）= LAPVIO
个体数（Individual number/15 hm^2）= 4
最大胸径（Max DBH）= 2.2 cm
重要值排序（Importance value rank）= 181

常绿灌木或半灌木，高 1~2 m。茎干时带紫褐色，上部与小枝生刺毛。叶宽卵形或近心形，被紧贴生的小刺毛和极稀疏的长刺毛，基出脉 3。花序雌雄同株，分枝短。瘦果倒卵形，歪斜，稍扁。花期 6~8 月，果期 8~11 月。

Evergreen shrubs or subshrubs, 1–2 m tall. Stems usually purplish, upper stem and branchlets with stinging hairs. Leaf blade broadly ovate ro cordate, densely appressed small stinging and sparsely long stinging, basal veins 3. Monoecism, shortly branched. Achene obovate, asymmetric. Fl. Jun.–Aug., fr. Aug.–Nov..

植株 Whole plant
摄影：黄俞淞 Photo by：Huang Yusong

叶 Leaves
摄影：黄俞淞 Photo by：Huang Yusong

雌花序 Female inflorescence
摄影：黄俞淞 Photo by：Huang Yusong

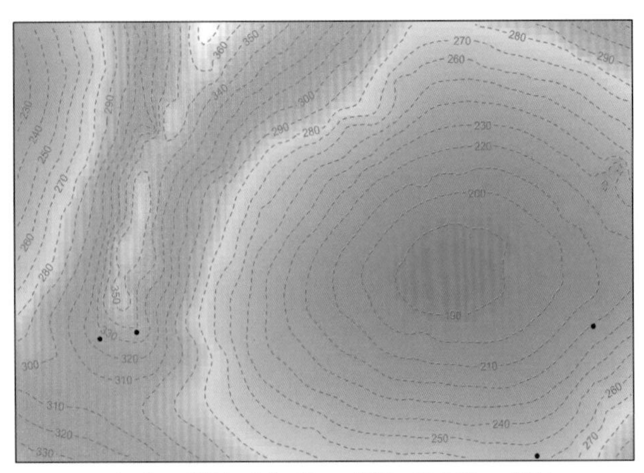

• 1~5 cm DBH　+ 5~20 cm DBH　○ ≥20 cm DBH
个体分布图 Distribution of individuals

径级分布表 DBH class

胸径区间 (Diameter class) (cm)	个体数 (No. of individuals in the plot)	比例 (Proportion) (%)
1~2	2	50.00
2~5	2	50.00
5~10	0	0.00
10~20	0	0.00
20~35	0	0.00
35~60	0	0.00
≥60	0	0.00

120 广西紫麻

guǎng xī zǐ má | Guangxi Woodnettle

Oreocnide kwangsiensis Hand.
荨麻科 Urticaceae

代码（SpCode）= OREKWA
个体数（Individual number/15 hm^2）= 20
最大胸径（Max DBH）= 5.2 cm
重要值排序（Importance value rank）= 134

灌木，高 1~1.5(3)m。叶坚纸质，狭椭圆形至椭圆状披针形，边缘全缘或在上部有极不明显的数枚圆齿，两面光滑，基出脉 3。果干时变黑色，圆锥形肉质花托壳斗状。花期 10 月至翌年 3 月，果期 5~10 月。

Shrubs, 1–1.5(3)m tall. Leaf blade stiffly papery, narrowly elliptic to elliptic-lanceolate, margin entire or sparesely and inconspicuously crenate-serrulate distally, both furfaces glabrous, basal veins 3. Achene black when dry, surrounded by a fleshy cupule. Fl. Oct.–Mar. of next year, fr. May–Oct..

花枝 Flowering branch
摄影：黄俞淞 Photo by：Huang Yusong

叶 Leaves
摄影：黄俞淞 Photo by：Huang Yusong

花序 Inflorescence
摄影：黄俞淞 Photo by：Huang Yusong

• 1~5 cm DBH + 5~20 cm DBH ○ ≥20 cm DBH
个体分布图 Distribution of individuals

径级分布表 DBH class

胸径区间 (Diameter class) (cm)	个体数 (No. of individuals in the plot)	比例 (Proportion) (%)
1~2	5	25.00
2~5	14	70.00
5~10	1	5.00
10~20	0	0.00
20~35	0	0.00
35~60	0	0.00
≥60	0	0.00

121 谷木叶冬青

gǔ mù yè dōng qīng | Memecylon-leaved Holly

Ilex memecylifolia Champ. ex Benth.
冬青科 Aquifoliaceae

代码（SpCode）= ILEMEM
个体数（Individual number/15 hm^2）= 1
最大胸径（Max DBH）= 1.4 cm
重要值排序（Importance value rank）= 215

常绿乔木，高可达 15~20 m。稀灌木，高仅 2 m。叶片革质至厚革质，卵状长圆形或倒卵形，全缘，两面无毛，托叶三角形，长约 0.5 mm，被微柔毛，宿存。花序簇生叶腋内。果球形，直径约 5~6 mm，成熟时红色，宿存柱头柱状。花期 3~4 月，果期 7~12 月.

Evergreen trees, up to 15–20 m tall. Rarely shrubs, ca. 2 m tall. Leaf blade leathery to stiffly leathery, ovate-oblong or obovate, margin entire, both surfaces glabrous, stipule trigonal, ca. 0.5 mm long, slightly pubescent, persistent. Inflorescences fascicled, axillary. Fruit globose, ca. 5–6 mm in diam., red when mature, persistent stigma columnar. Fl. Mar.–Apr., fr. Jul.–Dec..

枝叶 Branch and leaves
摄影：黄俞淞 Photo by: Huang Yusong

叶 Leaves
摄影：黄俞淞 Photo by: Huang Yusong

枝 Branch
摄影：黄俞淞 Photo by: Huang Yusong

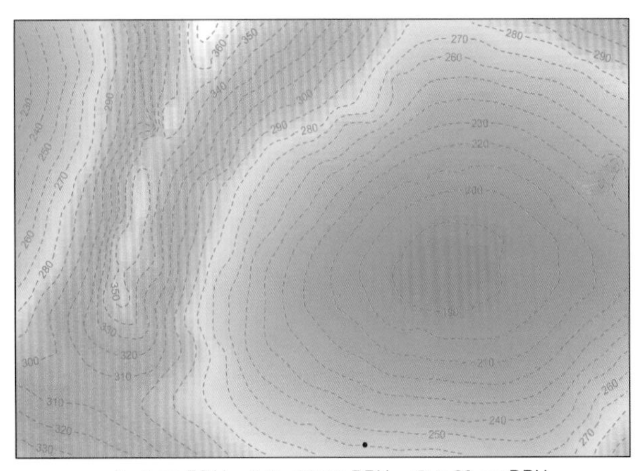

• 1~5 cm DBH + 5~20 cm DBH ○ ≥20 cm DBH
个体分布图 Distribution of individuals

径级分布表 DBH class

胸径区间 (Diameter class) (cm)	个体数 (No. of individuals in the plot)	比例 (Proportion) (%)
1~2	1	100.00
2~5	0	0.00
5~10	0	0.00
10~20	0	0.00
20~35	0	0.00
35~60	0	0.00
≥60	0	0.00

122 裂果卫矛

liè guǒ wèi máo | Diels's Spindle-tree

Euonymus dielsianus Loes. ex Diels
卫矛科 Celastraceae

代码（SpCode）= EUODIE
个体数（Individual number/15 hm^2）= 220
最大胸径（Max DBH）= 5.1 cm
重要值排序（Importance value rank）= 69

常绿灌木或小乔木, 高 1~7 m。叶革质，窄长椭圆形或长倒卵形。聚伞花序 1~7 花，蒴果 4 深裂，裂瓣卵形。花期 4~7 月，果期 9~11 月。

Evergreen shrubs or small trees, 1–7 m tall. Leaf blade leathery, narrowly oblong-elliptic or elliptic-obovate. Cymes 1–7-flowers, capsule 4-lobed, lobs ovate. Fl. Apr.–Jul., fr. Sep.–Nov..

枝叶 Branch and leaves
摄影：黄俞淞 Photo by：Huang Yusong

花 Flower
摄影：丁涛 Photo by：Ding Tao

果 Fruits
摄影：黄俞淞 Photo by：Huang Yusong

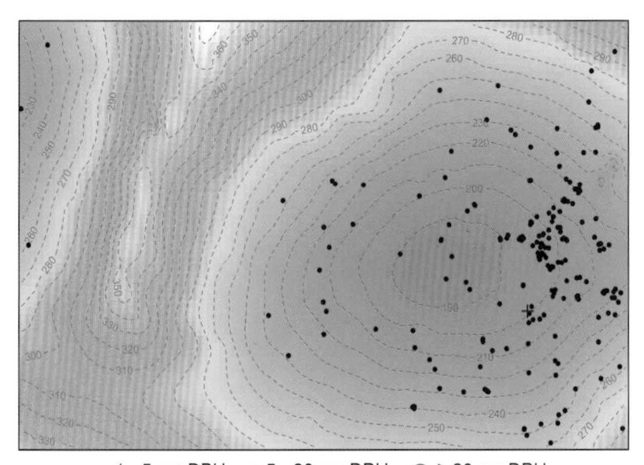

• 1~5 cm DBH + 5~20 cm DBH ○ ≥20 cm DBH
个体分布图 Distribution of individuals

径级分布表 DBH class

胸径区间 (Diameter class) (cm)	个体数 (No. of individuals in the plot)	比例 (Proportion) (%)
1~2	173	78.64
2~5	46	20.91
5~10	1	0.45
10~20	0	0.00
20~35	0	0.00
35~60	0	0.00
≥60	0	0.00

123 皱叶沟瓣

zòu yè gōu bàn | Wrinkly-leaved Glyptopetalum

Glyptopetalum rhytidophyllum (Chun et How) C. Y. Cheng
卫矛科 Celastraceae

代码（SpCode）= GLYRHY
个体数（Individual number/15 hm^2）= 1
最大胸径（Max DBH）= 1.1 cm
重要值排序（Importance value rank）= 216

常绿灌木，高 1.5~3 m，小枝绿色。叶薄革质，干时保持绿色，长方阔披针形，侧脉 8~18 对，在叶面下凹较深，叶面成皱缩状。花序 1~2 次分枝，小苞片锥形，长约 12 毫米，宿存。蒴果灰白色或淡棕色，外皮有糠秕状细斑块，圆球状。花期 6~8 月，果期 9~12 月。

Evergreen shrubs, 1.5–3 m tall. Branchlets green. Leaf blade thinly leathery, green when dry, narrowly oblong or broadly oblong-lanceolate,lateral veins 8–18 pairs, appearing rugose due to depressed veins. Inflorescences 1–2-branched, bractlets subulate, ca. 12 mm long, persistent. Capsule globose, pallid or slightly brown, squarrulose maculate. Fl. Jun.–Aug., fr. Sep.–Dec..

果皮和种子 Peel and seeds
摄影：丁涛 Photo by：Ding Tao

植株 Whole plant
摄影：黄俞淞 Photo by：Huang Yusong

果 Fruits
摄影：黄俞淞 Photo by：Huang Yusong

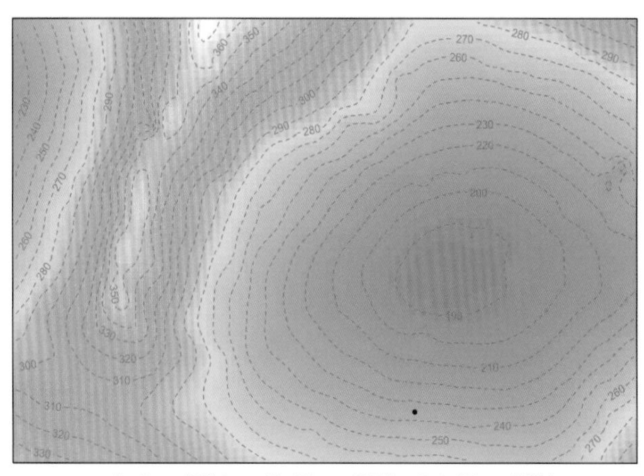

• 1~5 cm DBH　+ 5~20 cm DBH　○ ≥20 cm DBH
个体分布图 Distribution of individuals

径级分布表 DBH class

胸径区间 (Diameter class) (cm)	个体数 (No. of individuals in the plot)	比例 (Proportion) (%)
1~2	1	100.00
2~5	0	0.00
5~10	0	0.00
10~20	0	0.00
20~35	0	0.00
35~60	0	0.00
≥60	0	0.00

124 密花美登木

mǐ huā měi dēng mù | Crown-flowered Maytenus

Maytenus confertiflorus J. Y. Luo et X. X. Chen
卫矛科 Celastraceae

代码（SpCode）= MAYCON
个体数（Individual number/15 hm^2）= 24
最大胸径（Max DBH）= 5.1 cm
重要值排序（Importance value rank）= 140

常绿灌木，高达 4 m。小枝有刺。叶纸质，阔椭圆形或倒卵形，先端渐尖或有短尖头。聚伞花序多数集生叶腋，有花多至 60 朵，呈圆球形，花序梗极短或无。蒴果淡绿带紫色，三角球状。花期 9~10 月，果期 10~11 月。

Evergreen shrubs, up to 4 m tall. Branchlets spinosity. Leaf blade papery, broadly elliptic or obovate, apex auminate or with mucro. Cymes axillary, up to 60-flowered, globose, peduncle very short or without. Capsules slightly greenish purple, triangle-globose. Fl. Sep.–Oct., fr. Oct.–Nov..

花枝 Flowering branch
摄影：王斌 Photo by：Wang Bin

花 Flowers
摄影：王斌 Photo by：Wang Bin

果 Fruits
摄影：黄俞淞 Photo by：Huang Yusong

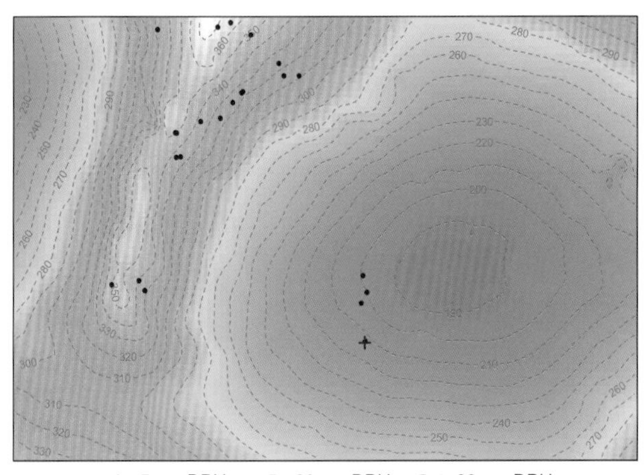

● 1~5 cm DBH ＋ 5~20 cm DBH ○ ≥20 cm DBH
个体分布图 Distribution of individuals

径级分布表 DBH class

胸径区间 (Diameter class) (cm)	个体数 (No. of individuals in the plot)	比例 (Proportion) (%)
1~2	19	79.17
2~5	4	16.67
5~10	1	4.17
10~20	0	0.00
20~35	0	0.00
35~60	0	0.00
≥60	0	0.00

125 柴龙树

chái lóng shù | Mugonyone

Apodytes dimidiata E. Mey. ex Arn.
茶茱萸科 Icacinaceae

代码（SpCode）= APODIM
个体数（Individual number/15 hm^2）= 235
最大胸径（Max DBH）= 31.7 cm
重要值排序（Importance value rank）= 30

灌木或乔木，高 3~10 m。树皮平滑，灰白色，小枝灰褐色。叶纸质，椭圆形或长椭圆形。圆锥花序顶生，密被黄色柔毛，花淡黄色或白色。核果长圆形，成熟前青色，成熟后红至黑色，基部有一盘状附属物，其一侧为宿存花柱。花果期全年。

Shrubs or trees, 3–10 m tall. Bark glabrous, grayish white, branchlets grayish brown. Leaf blade papery, elliptic or oblong-elliptic. Panicles terminal, densely yellowish pubescent, flowers yellowish or white. Drupe oblong, green when young, red to black-red when mature, base with a discoid fleshy appendage, with persistent style. Fl. and fr. all seasons.

树干 Trunk
摄影：王斌 Photo by: Wang Bin

果枝 Fruiting branch
摄影：黄俞淞 Photo by: Huang Yusong

果 Fruits
摄影：黄俞淞 Photo by: Huang Yusong

• 1~5 cm DBH + 5~20 cm DBH ○ ≥20 cm DBH
个体分布图 Distribution of individuals

径级分布表 DBH class

胸径区间 (Diameter class) (cm)	个体数 (No. of individuals in the plot)	比例 (Proportion) (%)
1~2	32	13.62
2~5	71	30.21
5~10	64	27.23
10~20	56	23.83
20~35	12	5.11
35~60	0	0.00
≥60	0	0.00

126 粗丝木

cū sī mù | Four-stamen Gomphandra

Gomphandra tetrandra (Wall.) Sleum.
茶茱萸科 Icacinaceae

代码（SpCode）＝ GOMTET
个体数（Individual number/15 hm^2）＝ 62
最大胸径（Max DBH）＝ 4.0 cm
重要值排序（Importance value rank）＝ 128

灌木或小乔木，高 2~10 m。树皮灰色，嫩枝绿色。叶纸质，狭披针形、长椭圆形或阔椭圆形，侧脉斜上升。聚伞花序与叶对生，少腋生，雄花黄白色或白绿色，雌花黄白色。核果椭圆形，成熟时白色，浆果状。花果期全年。

Shrubs or small trees, 2–10 m tall. Bark gray, branchlets green. Leaf blade papery, narrowly elliptic or broadly elliptic, lateral veins obliquely ascending. Cymes opposite leaves, rarely axillary, male flowers yellowish white or whitish green, female flowers yellowish white. Drupes elliptic, white when mature, berrylike. Fl. and fr. all seasons.

叶 Leaves
摄影：黄俞淞 Photo by：Huang Yusong

果序 Infructescence
摄影：黄俞淞 Photo by：Huang Yusong

花序 Inflorescence
摄影：黄俞淞 Photo by：Huang Yusong

• 1~5 cm DBH　+ 5~20 cm DBH　○ ≥20 cm DBH
个体分布图 Distribution of individuals

径级分布表 DBH class

胸径区间 (Diameter class) (cm)	个体数 (No. of individuals in the plot)	比例 (Proportion) (%)
1~2	47	75.81
2~5	15	24.19
5~10	0	0.00
10~20	0	0.00
20~35	0	0.00
35~60	0	0.00
≥60	0	0.00

127 茎花山柚

jìng huā shān yòu | Long-stamen Champereia

Champereia manillana var. *longistaminea* (W. Z. Li) H. S. Kiu
山柚子科 Opiliaceae

代码（SpCode）= CHAMAN
个体数（Individual number/15 hm^2）= 1340
最大胸径（Max DBH）= 21.6 cm
重要值排序（Importance value rank）= 13

常绿小乔木，高 6~7 m。树皮青白色，无毛。叶革质，易折断，全缘，干后苍白色。圆锥花序状聚伞花序，腋生。核果椭圆形，无毛，橙红色。花期 4~6 月，果期 6~10 月。

Evergreen small trees, 6–7 m tall. Bark pale, glabrous. Leaf blade leathery, break off easyly, entire, pale when dry. Panicles cymose, axillary. Drupes elliptic, glabrous, orange-red. Fl. Apr.–Jun., fr. Jun.–Oct..

果序 Infructescence
摄影：黄俞淞 Photo by：Huang Yusong

花序 Inflorescence
摄影：黄俞淞 Photo by：Huang Yusong

叶 Leaves
摄影：黄俞淞 Photo by：Huang Yusong

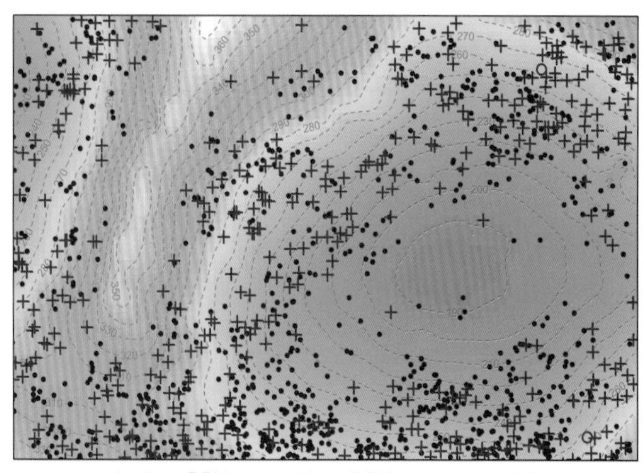

• 1~5 cm DBH + 5~20 cm DBH ○ ≥20 cm DBH
个体分布图 Distribution of individuals

径级分布表 DBH class

胸径区间 (Diameter class) (cm)	个体数 (No. of individuals in the plot)	比例 (Proportion) (%)
1~2	387	28.88
2~5	522	38.96
5~10	321	23.96
10~20	108	8.06
20~35	2	0.15
35~60	0	0.00
≥60	0	0.00

128 革叶鼠李

gé yè shǔ lǐ | Leather-leaf Buckthorn

Rhamnus coriophylla Hand.-Mazz.
鼠李科 Rhamnaceae

代码（SpCode）= RHACOR
个体数（Individual number/15 hm^2）= 26
最大胸径（Max DBH）= 4.4 cm
重要值排序（Importance value rank）= 131

常绿灌木或小乔木，高 3~4 m。小枝紫褐色，具多数瘤状皮孔。叶革质，椭圆形或矩圆状椭圆形，边缘干时常背卷，叶背近在脉腋有髯毛。腋生聚伞总状花序，长 1~3 cm。单性，雌雄异株。核果倒卵状球形，基部有宿存的萼筒，成熟时紫红色。花期 6~8 月，果期 8~12 月。

Evergreen shrubs or small trees, 3–4 m tall. Branchlets purple-brown, with numerous tuberculate lenticels. Leaf blade leathery, elliptic or oblong-elliptic, margin often revolute when dry, abaxially barbellate in axils of veins only. Cymes axillary, 1–3 cm long, solitary, dioecious. Drupes obovoid-globose, with persistent calyx tube at base, purle-red when mature. Fl. Jun.–Aug., fr. Aug.–Dec..

果枝 Fruiting branches
摄影：黄俞淞 Photo by：Huang Yusong

叶 Leaves
摄影：黄俞淞 Photo by：Huang Yusong

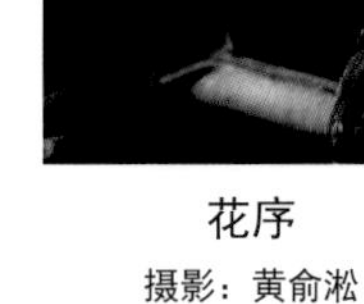

花序 Inflorescence
摄影：黄俞淞 Photo by：Huang Yusong

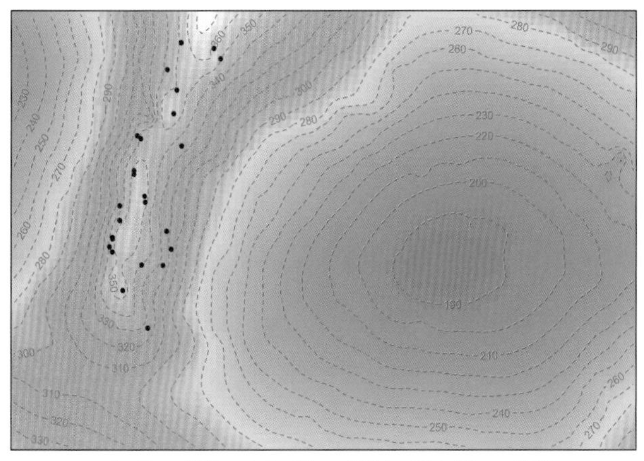

• 1~5 cm DBH + 5~20 cm DBH ○ ≥20 cm DBH
个体分布图 Distribution of individuals

径级分布表 DBH class

胸径区间 (Diameter class) (cm)	个体数 (No. of individuals in the plot)	比例 (Proportion) (%)
1~2	15	57.69
2~5	11	42.31
5~10	0	0.00
10~20	0	0.00
20~35	0	0.00
35~60	0	0.00
≥60	0	0.00

129 印度枣

yìng dù zǎo | Indian Jujube

Ziziphus incurva Roxb.
鼠李科 Rhamnaceae

代码（SpCode）= ZIZINC
个体数（Individual number/15 hm^2）= 86
最大胸径（Max DBH）= 15.4 cm
重要值排序（Importance value rank）= 85

落叶乔木，高达 15 m。幼枝被棕色短柔毛，具皮刺。叶纸质，卵状矩圆形或卵形，基部近圆形，边缘具圆锯齿，基出脉 3~5。花绿色，数个至十余个密集成腋生二歧式聚伞花序。核果近球形或球状椭圆形，无毛，基部有宿存的萼筒，成熟时红褐色。花期 4~5 月，果期 6~10 月。

Deciduous trees, up to 15 m tall. Young branches brownish pilose, spinose. Leaf blade papery, ovate-oblong or ovate, base rounded, margin crenate-serrate, basal veins 3–5. Flowers greenish, several to 10 in axillary dichotomous cymes. Drupe subglobose or globose-ellipsoid, glabrous, with persistent calyx tube at base, red-brown when mature. Fl. Apr.–May, fr. Jun.–Oct..

枝叶 Branch and leaves
摄影：黄俞淞 Photo by: Huang Yusong

叶 Leaf
摄影：黄俞淞 Photo by: Huang Yusong

果 Fruits
摄影：赖阳均 Photo by: Lai Yangjun

• 1~5 cm DBH + 5~20 cm DBH ○ ≥20 cm DBH
个体分布图 Distribution of individuals

径级分布表 DBH class

胸径区间 (Diameter class) (cm)	个体数 (No. of individuals in the plot)	比例 (Proportion) (%)
1~2	27	31.40
2~5	32	37.21
5~10	25	29.07
10~20	2	2.33
20~35	0	0.00
35~60	0	0.00
≥60	0	0.00

130 火筒树

huǒ tǒng shù | Indian Leea

Leea indica (Burm. f.) Merr.
葡萄科 Vitaceae

代码（SpCode）= LEEIND
个体数（Individual number/15 hm^2）= 185
最大胸径（Max DBH）= 17.2 cm
重要值排序（Importance value rank）= 70

常绿直立灌木，高 1.5~4 m。小枝圆柱形，具纵棱。叶二至三回羽状复叶，边缘有不整齐锯齿。花序与叶对生，复二歧聚伞花序或二级分枝集生成伞形，总花梗被褐色柔毛。果实扁球形，有种子 4~6 颗。花期 4~7 月，果期 8~12 月。

Evergreen erect shrubs, 1.5–4 m tall. Branchlets terete, with longitudinal ridges. Leaves 2–3-pinnate, margin wiht irregular teeth. Inflorescences opposite to leaves, compound dichasial or umbelliform, peduncle with brown hairs. Berry compressed globose, 4–6-seeded. Fl. Apr.–Jul., fr. Aug.–Dec..

植株 Whole plant
摄影：黄俞淞 Photo by：Huang Yusong

花序 Inflorescence
摄影：黄俞淞 Photo by：Huang Yusong

果 Fruits
摄影：黄俞淞 Photo by：Huang Yusong

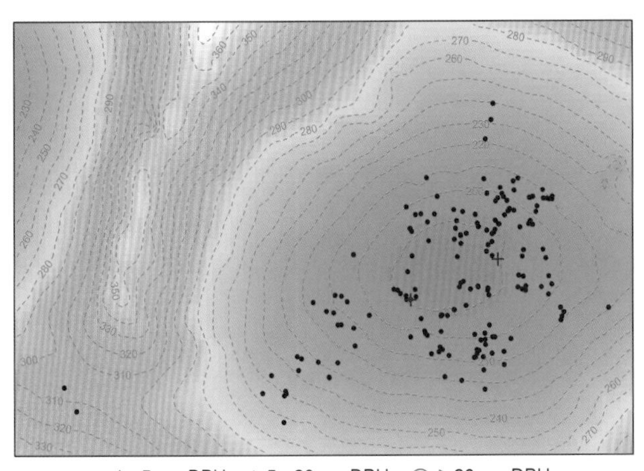

• 1~5 cm DBH + 5~20 cm DBH ○ ≥20 cm DBH
个体分布图 Distribution of individuals

径级分布表 DBH class

胸径区间 (Diameter class) (cm)	个体数 (No. of individuals in the plot)	比例 (Proportion) (%)
1~2	64	34.59
2~5	119	64.32
5~10	1	0.54
10~20	1	0.54
20~35	0	0.00
35~60	0	0.00
≥60	0	0.00

131 柚

yòu | Shaddock

Citrus maxima (Burm.) Merr.
芸香科 Rutaceae

代码（SpCode）= CITMAX
个体数（Individual number/15 hm^2）= 1
最大胸径（Max DBH）= 4.2 cm
重要值排序（Importance value rank）= 206

常绿乔木。嫩枝通常鲜红色，扁且有棱。叶阔卵形或椭圆形，基部圆，顶端圆或钝，或有时具短尖，叶柄具翼。总状花序，腋生。果横径通常 10 cm 以上，油胞大，凸起。花期 4~5 月，果期 9~12 月。

Evergreen trees. Branches usually purplish, flat with ridges when young. Leaf blade broadly ovate or elliptic, base rounded, apex rounded to obtuse, or sometimes mucronate, petiole winged. Flowers solitary or in racemes. Fruit usually more than 10 cm in diam., with large prominent oil dots. Fl. Apr.–May, fr. Sep. Dec..

果枝 Fruiting branches
摄影：许为斌 Photo by：Xu Weibin

叶 Leaves
摄影：王斌 Photo by：Wang Bin

花 Flowers
摄影：许为斌 Photo by：Xu Weibin

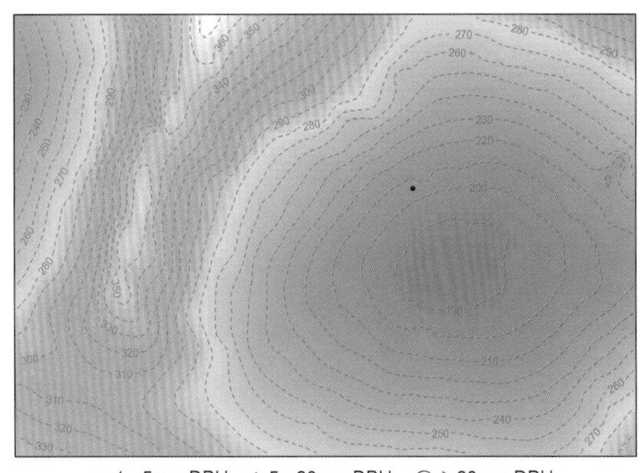

• 1~5 cm DBH + 5~20 cm DBH ○ ≥20 cm DBH
个体分布图 Distribution of individuals

径级分布表 DBH class

胸径区间 (Diameter class) (cm)	个体数 (No. of individuals in the plot)	比例 (Proportion) (%)
1~2	0	0.00
2~5	1	100.00
5~10	0	0.00
10~20	0	0.00
20~35	0	0.00
35~60	0	0.00
≥60	0	0.00

132 细叶黄皮

xì yè huáng pí | Sanki Wampee

Clausena anisum-olens (Blanco) Merr.
芸香科 Rutaceae

代码（SpCode）= CLAANI
个体数（Individual number/15 hm^2）= 176
最大胸径（Max DBH）= 27.7 cm
重要值排序（Importance value rank）= 64

常绿小乔木，高 3~6 m。各部密生透明油点。叶有小叶 5~11 片，小叶镰刀状披针形或斜卵形，两侧不对称。花序顶生，花白色。果圆球形，淡黄色，或淡朱红色，果皮有多数肉眼可见的半透明油点。花期 4~5 月，果期 7~8 月。

Evergreen small trees, 3–6 m tall. Densely transparent oil drop. Leaflets 5–11, sickle shaped, lanceolate or ovate, base asymmetric. Inflorescences terminal, flowers white. Fruit globose, slightly yellow, or slightly imperial red, peel with conspicuous subtransparent oil dots. Fl. Apr.–May, fr. Jul.–Aug..

树干 Trunk
摄影：黄俞淞 Photo by：Huang Yusong

枝叶 Branch and leaves
摄影：黄俞淞 Photo by：Huang Yusong

果序 Infructescence
摄影：黄俞淞 Photo by：Huang Yusong

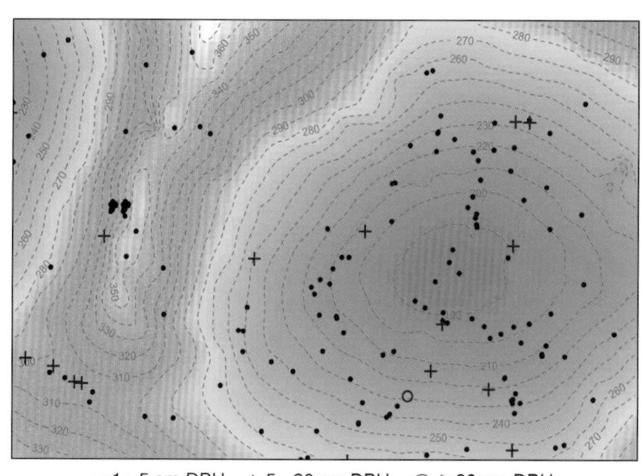

• 1~5 cm DBH　+ 5~20 cm DBH　○ ≥20 cm DBH
个体分布图 Distribution of individuals

径级分布表 DBH class

胸径区间 (Diameter class) (cm)	个体数 (No. of individuals in the plot)	比例 (Proportion) (%)
1~2	81	46.02
2~5	77	43.75
5~10	12	6.82
10~20	5	2.84
20~35	1	0.57
35~60	0	0.00
≥60	0	0.00

133 齿叶黄皮

chǐ yè huáng pí | Dunn's Wampee

Clausena dunniana H. Lév.
芸香科 Rutaceae

代码（SpCode）= CLADUN
个体数（Individual number/15 hm^2）= 394
最大胸径（Max DBH）= 19.1 cm
重要值排序（Importance value rank）= 48

落叶小乔木，高 2~5 m。叶有小叶 5~15 片，小叶卵形至披针形，基部两侧不对称。花序顶生兼有生于小枝的近顶部叶腋间，花梗无毛；果近圆球形，初时暗黄色，后变红色，透熟时紫黑色。花期 6~7 月，果期 10~11 月。

Deciduous small trees, 2–5 m tall.Leaves 5–15-foliolate, leaflet blades ovate to lanceolate, base asymmetric. Inflorescences terminal, pedicel glabrous. Fruit globose, grey-yellow to red or purple black. Fl. Jun.–Jul., fr. Oct.–Nov..

树干 Trunk
摄影：王斌 Photo by：Wang Bin

果序 Infructescence
摄影：刘晟源 Photo by：Liu Shengyuan

复叶 Compound leaf
摄影：王斌 Photo by：Wang Bin

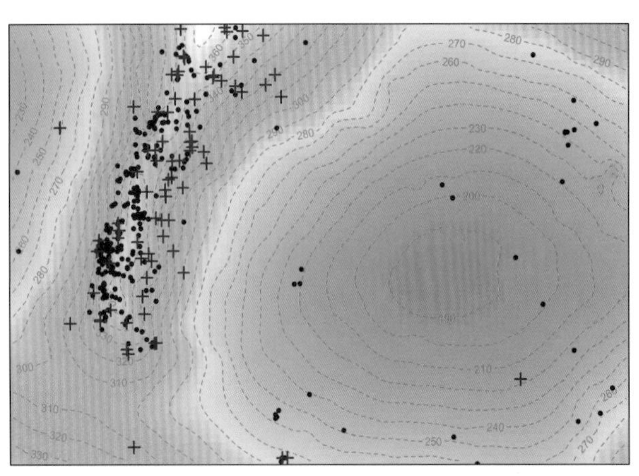

• 1~5 cm DBH + 5~20 cm DBH ○ ⩾20 cm DBH
个体分布图 Distribution of individuals

径级分布表 DBH class

胸径区间 (Diameter class) (cm)	个体数 (No. of individuals in the plot)	比例 (Proportion) (%)
1~2	162	41.12
2~5	153	38.83
5~10	54	13.71
10~20	25	6.35
20~35	0	0.00
35~60	0	0.00
⩾60	0	0.00

134 锈毛山小橘

xiù máo shān xiǎo jú | Esquirol's Glycosmis

Glycosmis esquirolii (Lévl.) Tanaka
芸香科 Rutaceae

代码（SpCode）= GLYESQ
个体数（Individual number/15 hm^2）= 11
最大胸径（Max DBH）= 6.4 cm
重要值排序（Importance value rank）= 171

小乔木，高 6~10 m。叶为复叶，有小叶 4~7 片，长椭圆形，长 10~16 cm，基部楔形，边缘具锯齿，顶部渐尖，钝头，圆锥花序，顶生及腋生，萼裂片阔卵形，长约 1 mm；花瓣淡黄白色，长 3~4 mm，子房近圆球形，与花柱同被微柔毛，花柱甚短。花期 10 月至翌年 3 月，果期 4 月。

Small trees, 6–10 m tall. Leaves 4–7-foliolate; leaflet blades oblong, 10–16 cm long, base cuneate, margin dentate, apex acuminate to obtuse. Inflorescences axillary or terminal, paniculate. Sepals broadly ovate, ca. 1 mm. Petals pale yellowish white, 3–4 mm. Ovary subglobose, rust-colored villosulous; style extremely short. Fl. Oct.–Mar. of next year, fr. Apr..

花枝 Flowering branch
摄影：黄俞淞 Photo by：Huang Yusong

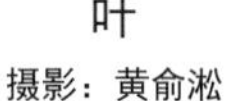

叶 Leaves
摄影：黄俞淞 Photo by：Huang Yusong

花序 Inflorescence
摄影：黄俞淞 Photo by：Huang Yusong

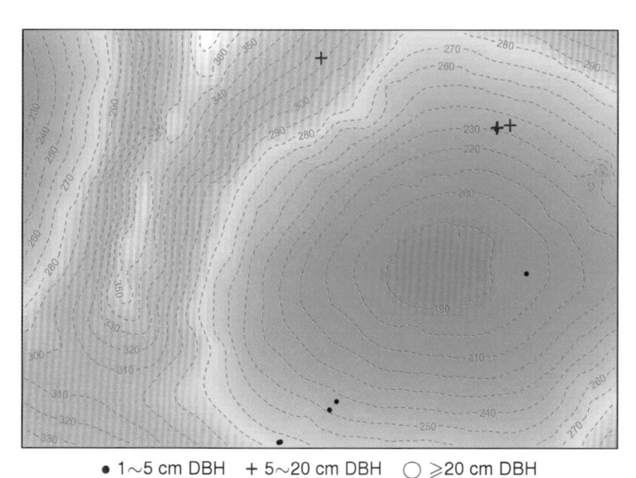

• 1~5 cm DBH + 5~20 cm DBH ○ ≥20 cm DBH
个体分布图 Distribution of individuals

径级分布表 DBH class

胸径区间 (Diameter class) (cm)	个体数 (No. of individuals in the plot)	比例 (Proportion) (%)
1~2	1	9.09
2~5	6	54.55
5~10	4	36.36
10~20	0	0.00
20~35	0	0.00
35~60	0	0.00
≥60	0	0.00

135 小芸木

xiǎo yún mù | Entire Micromelum

Micromelum integerrimum (Buch. -Ham.) Roem.
芸香科 Rutaceae

代码（SpCode）= MICINT
个体数（Individual number/15 hm^2）= 7
最大胸径（Max DBH）= 7.5 cm
重要值排序（Importance value rank）= 170

灌木或乔木，高 3~12 m。小枝灰褐色。叶互生，叶片椭圆形或长卵形，基部短狭或楔形，背面近缘常有数个圆形的腺体，叶柄长 2~7.5 cm，顶端具 2 毗连的腺体。花单性，雌雄同株，密集成长 4~9 cm 的顶生总状花序。蒴果黑色，球形。花期 4~6 月，果期 7~10 月。

Shrubs or trees, 3–12 m tall. Branchlets gray-brown. Leaves alternate, elliptic or oblong-ovate, base cuneate, with several rounded glands on or near margin abaxially, petioles 2–7.5 cm, 2-glandular at apex. Flowers monoecious in terminal racemes, inflorescences 4–9 cm long. Capsules black, globose. Fl. Apr.–Jun., fr. Jul.–Oct..

植株 Whole plant
摄影：黄俞淞 Photo by：Huang Yusong

复叶 Compound leaf
摄影：黄俞淞 Photo by：Huang Yusong

果 Fruits
摄影：黄俞淞 Photo by：Huang Yusong

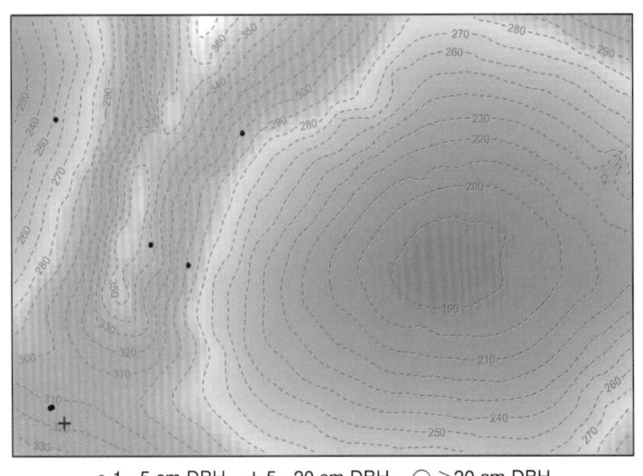

• 1~5 cm DBH + 5~20 cm DBH ○ ≥20 cm DBH
个体分布图 Distribution of individuals

径级分布表 DBH class

胸径区间 (Diameter class) (cm)	个体数 (No. of individuals in the plot)	比例 (Proportion) (%)
1~2	2	28.57
2~5	4	57.14
5~10	1	14.29
10~20	0	0.00
20~35	0	0.00
35~60	0	0.00
≥60	0	0.00

136 楝叶吴萸

liàn yè wú yú | Glabrous Euodia

Tetradium glabrifolium (Champ. ex Benth.) Hartley
芸香科 Rutaceae

代码（SpCode）= TETGLA
个体数（Individual number/15 hm^2）= 13
最大胸径（Max DBH）= 18.4 cm
重要值排序（Importance value rank）= 143

乔木，高达 20 m。树皮灰白色。叶有小叶 3~19，小叶阔卵状披针形，两则明显不对称，叶缘有细钝齿或全缘，无毛。花序顶生，花瓣白色。分果瓣淡紫红色，径约 5 mm，有成熟种子 1 粒，褐黑色。花期 7~9 月，果期 10~12 月。

Trees, up to 20 m tall. Bark grayish white. Leaves 3–19-foliolated, leaflet blade broadly ovate to lanceolate, inequilateral, margin entire or slightly crenulate, glabrous. Inflorescences terminal, petal white. Follicles purplish red, ca. 5 mm in diam., with 1 mature seed, brownish black. Fl. Jul.–Sep., fr. Oct.–Dec..

树干 Trunk
摄影：黄俞淞 Photo by: Huang Yusong

果枝 Fruiting branches
摄影：黄俞淞 Photo by: Huang Yusong

枝叶 Branch and leaves
摄影：黄俞淞 Photo by: Huang Yusong

• 1~5 cm DBH + 5~20 cm DBH ○ ≥20 cm DBH
个体分布图 Distribution of individuals

径级分布表 DBH class

胸径区间 (Diameter class) (cm)	个体数 (No. of individuals in the plot)	比例 (Proportion) (%)
1~2	3	23.08
2~5	1	7.69
5~10	3	23.08
10~20	6	46.15
20~35	0	0.00
35~60	0	0.00
≥60	0	0.00

137 簕欓花椒 lè dǎng huā jiāo | Avicenna's Prickly-ash

Zanthoxylum avicennae (Lam.) DC.
芸香科 Rutaceae

代码（SpCode）= ZANAVI
个体数（Individual number/15 hm^2）= 1
最大胸径（Max DBH）= 11.4 cm
重要值排序（Importance value rank）= 204

落叶乔木，高达 15 m。树干具刺，各部无毛。叶有小叶 11~21 片，小叶通常对生，斜卵形，斜长方形或呈镰刀状，两侧甚不对称，全缘，或中部以上有疏裂齿。花序顶生，花序轴及花梗有时紫红色。分果瓣淡紫红色，单个分果瓣径 4~5 mm，顶端无芒尖。花期 6~8（10）月，果期 10~12 月。

Deciduous trees, up to 15 m tall. Trunk with prickles, glabrous. Leaves 11–21-foliolate, leaflets usually opposite, obliquely ovate, obliquely oblong or falcate, inequilateral, entire, or apically crenate. Inflorescences teminal, rachis and pedicel sometimes purple red. Follicles 4–5 mm in diam., apex not beaked. Fl. Jun.–Aug. (Oct.), fr. Oct.–Dec..

树干 Trunk
摄影：丁涛 Photo by：Ding Tao

枝叶 Branch and leaves
摄影：丁涛 Photo by：Ding Tao

花序 Inflorescence
摄影：黄俞淞 Photo by：Huang Yusong

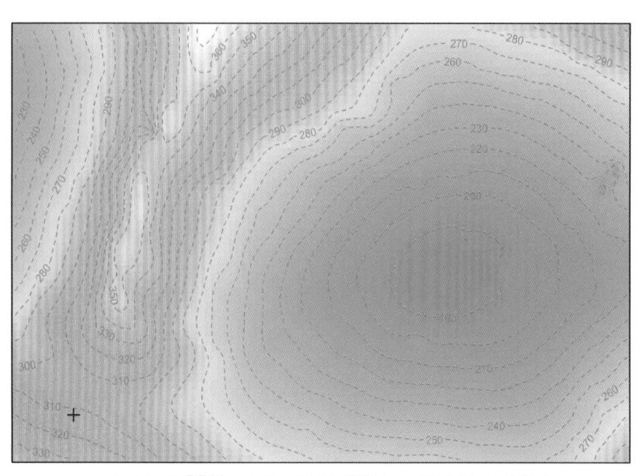

• 1~5 cm DBH + 5~20 cm DBH ○ ≥20 cm DBH
个体分布图 Distribution of individuals

径级分布表 DBH class

胸径区间 (Diameter class) (cm)	个体数 (No. of individuals in the plot)	比例 (Proportion) (%)
1~2	0	0.00
2~5	0	0.00
5~10	0	0.00
10~20	1	100.00
20~35	0	0.00
35~60	0	0.00
≥60	0	0.00

138 苦树

kǔ shù | Picrasma

Picrasma quassioides (D. Don) Benn.
苦木科 Simaroubaceae

代码（SpCode）= PICQUA
个体数（Individual number/15 hm^2）= 4
最大胸径（Max DBH）= 12.3 cm
重要值排序（Importance value rank）= 180

落叶乔木，高达 10 m。树皮紫褐色，平滑，有灰色斑纹。奇数羽状复叶，边缘具不整齐的粗锯齿，叶面无毛，落叶后留有明显的半圆形或圆形叶痕。花雌雄异株，组成腋生复聚伞花序。核果成熟后蓝绿色，萼宿存。花期 4~5 月，果期 6~9 月。

Deciduous trees, up to 10 m tall. Bark purplish brown, smooth, with gray stripes. Leaves altemate, odd-pinnate, margin irregularly serrate, leaf surfaces glabrous, leaf scar consipcuous, semirounded or rounded. Flowers dioecious, in axillary cymes. Drupes blue-green when mature, calyx persistent. Fl. Apr.–May, fr. Jun.–Sep..

果枝 Fruiting branches
摄影：黄俞淞 Photo by: Huang Yusong

核果 Drupes
摄影：黄俞淞 Photo by: Huang Yusong

复叶 Compound leaves
摄影：黄俞淞 Photo by: Huang Yusong

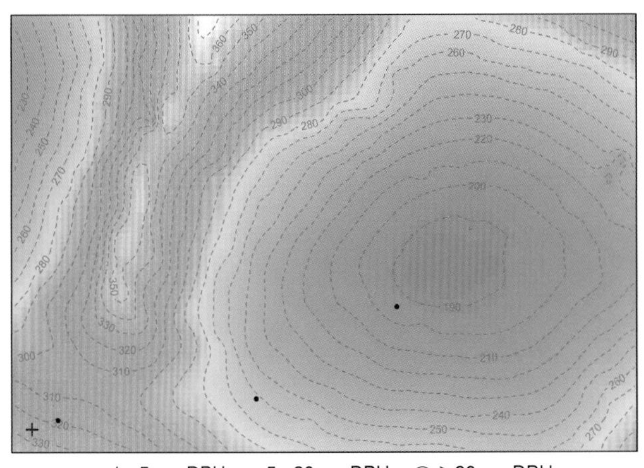

• 1~5 cm DBH + 5~20 cm DBH ○ ≥20 cm DBH
个体分布图 Distribution of individuals

径级分布表 DBH class

胸径区间 (Diameter class) (cm)	个体数 (No. of individuals in the plot)	比例 (Proportion) (%)
1~2	2	50.00
2~5	1	25.00
5~10	0	0.00
10~20	1	25.00
20~35	0	0.00
35~60	0	0.00
≥60	0	0.00

139 白头树

bái tóu shù | Forrest's Garuga

Garuga forrestii W. W. Smith
橄榄科 Burseraceae

代码（SpCode）= GARFOR
个体数（Individual number/15 hm^2）= 10
最大胸径（Max DBH）= 101.0 cm
重要值排序（Importance value rank）= 87

落叶乔木，高 10~15(25)m。幼枝密被柔毛，老枝无毛，有纵条纹及明显的叶痕。叶有小叶 11~19，披针形至椭圆状长圆形，幼时密被柔毛。圆锥花序侧生和腋生。果近卵形，两端尖，先端具喙而偏斜一侧，基部有宿存的浅杯状花萼。花期 4 月，果期 5~11 月。

Deciduous trees, 10–15(25) m tall. Branchlets densely pubescent when young, glabrescent, with longitudinally striped and conspicuously lenticellate. Leaflets 11–19, lanceolate to elliptic oblong, densely pubescent when young. Panicles axillary or lateral. Drupe nearly ovoid, attenuate at both ends, apex with a point, base with persistent shallowly cupular calyx. Fl. Apr., fr. May–Nov..

树干 Trunk
摄影：王斌 Photo by：Wang Bin

幼苗 Seedling
摄影：张金龙 Photo by：Zhang Jinlong

复叶 Compound leaf
摄影：张金龙 Photo by：Zhang Jinlong

• 1~5 cm DBH + 5~20 cm DBH ○ ≥20 cm DBH
个体分布图 Distribution of individuals

径级分布表 DBH class

胸径区间 (Diameter class) (cm)	个体数 (No. of individuals in the plot)	比例 (Proportion) (%)
1~2	0	0.00
2~5	0	0.00
5~10	0	0.00
10~20	0	0.00
20~35	4	40.00
35~60	5	50.00
≥60	1	10.00

140 羽叶白头树

yǔ yè bái tóu shù | Pinnate-leaf Garuga

Garuga pinnata Roxb.
橄榄科 Burseraceae

代码（SpCode）= GARPIN
个体数（Individual number/15 hm^2）= 46
最大胸径（Max DBH）= 74.0 cm
重要值排序（Importance value rank）= 51

落叶乔木，高可达 20 m。树皮灰褐色，粗糙，有皮孔和明显的叶痕。叶有小叶 9~23 片，小叶两面被长柔毛，长圆形或披针形，叶基部偏斜。圆锥花序腋生和侧生，幼时密被粗长柔毛。果近球形，成熟时黄色，有时被柔毛。花期 3~4 月，果期 4~10 月。

Deciduous trees, to 20 m tall. Bark grayish brown, asperous, with lenticel and obvious leaf scar. Leaflets 9–23, villous on both faces, oblong or lanceolate, base oblique. Panicle axillary and lateral, confert villose when infancy; fruit suborbicular, yellow at maturitye, sometimes pubescent. Fl. Mar.–Apr., fr. Apr.–Oct..

树干 Trunk
摄影：王斌 Photo by：Wang Bin

果枝 Fruiting branches
摄影：刘晟源 Photo by：Liu Shengyuan

果序 Infructescence
摄影：刘晟源 Photo by：Liu Shengyuan

• 1~5 cm DBH + 5~20 cm DBH ○ ≥20 cm DBH
个体分布图 Distribution of individuals

径级分布表 DBH class

胸径区间 (Diameter class) (cm)	个体数 (No. of individuals in the plot)	比例 (Proportion) (%)
1~2	0	0.00
2~5	4	8.70
5~10	6	13.04
10~20	6	13.04
20~35	17	36.96
35~60	10	21.74
≥60	3	6.52

141 望谟崖摩

wàng mó yá mó | Law's Amoora

Aglaia lawii (Wight) C. J. Saldanha et Ramamorthy
楝科 Meliaceae

代码（SpCode）= AGLLAW
个体数（Individual number/15 hm^2）= 95
最大胸径（Max DBH）= 21.6 cm
重要值排序（Importance value rank）= 83

灌木或乔木，高可达 20 m。叶互生，叶为奇数或偶数羽状复叶，小叶片椭圆形，长圆形，卵状披针形或披针形，叶基部不对称。圆锥花序腋生，远短于叶。蒴果球形，被鳞片。花期 5~12 月，果期几乎全年。

Shrubs or trees, up to 20 m tall. Leaves alternate, odd or even-pinnate, leaflet blades elliptic, oblong, ovate-lanceolate or lanceolate, base inequilateral. Thyrses axillary, usually shorter than leaves. Capsules globose, lepidote. Fl. May–Dec., fr. almost seasons.

树干 Trunk
摄影：王斌 Photo by: Wang Bin

果 Fruits
摄影：黄俞淞 Photo by: Huang Yusong

复叶 Compound leaves
摄影：丁涛 Photo by: Ding Tao

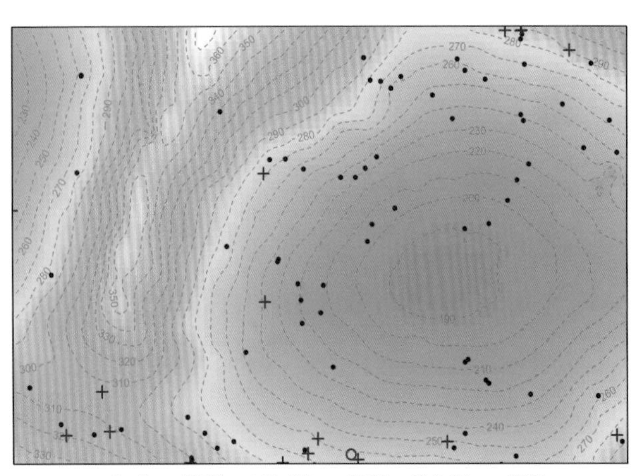

• 1~5 cm DBH + 5~20 cm DBH ○ ≥20 cm DBH
个体分布图 Distribution of individuals

径级分布表 DBH class

胸径区间 (Diameter class) (cm)	个体数 (No. of individuals in the plot)	比例 (Proportion) (%)
1~2	61	64.21
2~5	16	16.84
5~10	10	10.53
10~20	7	7.37
20~35	1	1.05
35~60	0	0.00
≥60	0	0.00

142 弄岗米仔兰

nòng gǎng mǐ zǎi lán | Longgang Aglaia

Aglaia species No.1
楝科 Meliaceae

代码（SpCode）= AGLSP1
个体数（Individual number/15 hm^2）= 787
最大胸径（Max DBH）= 25.7 cm
重要值排序（Importance value rank）= 26

灌木或小乔木。幼枝顶部被星状锈色的鳞片。叶轴和叶柄具狭翅，有小叶 3~7(9) 片，小叶对生，顶端 1 片最大。圆锥花序腋生，长 5~10 cm。果为浆果，初时被散生的星状鳞片，后脱落。花期 5~12 月，果期 7 月至翌年 3 月。

Shrubs or small trees. Young branches apically with stellate or lepidote trichomes. Petiole and rachis narrowly winged, leaflets opposite, 3–7(9), the apical one largest. Thyrses axillary, 5–10 cm long. Berry with stellate lepidote when young, glabrescent. Fl. May–Dec., fr. Jul.–Mar. of next year.

树干 Trunk
摄影：王斌 Photo by：Wang Bin

浆果 Berries
摄影：黄俞淞 Photo by：Huang Yusong

复叶 Compound leaves
摄影：黄俞淞 Photo by：Huang Yusong

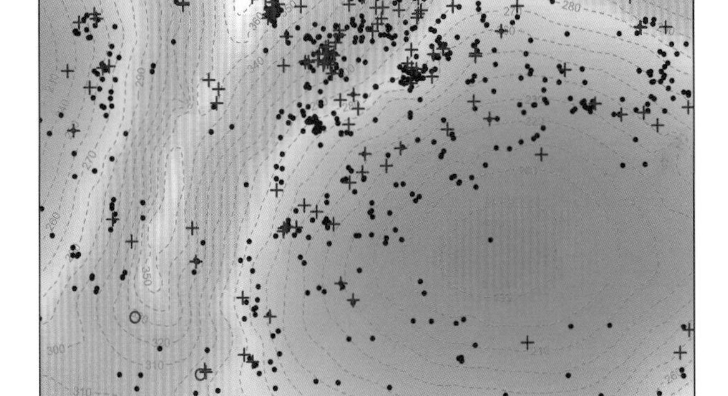

• 1~5 cm DBH + 5~20 cm DBH ○ ≥20 cm DBH
个体分布图 Distribution of individuals

径级分布表 DBH class

胸径区间 (Diameter class) (cm)	个体数 (No. of individuals in the plot)	比例 (Proportion) (%)
1~2	406	51.59
2~5	246	31.26
5~10	89	11.31
10~20	43	5.46
20~35	3	0.38
35~60	0	0.00
≥60	0	0.00

143 山楝

shān liàn | Manystamen Aphanamixis

Aphanamixis polystachya (Wall.) R. Parker
楝科 Meliaceae

代码（SpCode）= APHPOL
个体数（Individual number/15 hm^2）= 12
最大胸径（Max DBH）= 23.9 cm
重要值排序（Importance value rank）= 148

乔木，高可达 20~30 m。叶为奇数羽状复叶，有小叶 9~11 (15) 片，小叶对生，长椭圆形，基部楔形或宽楔形，偏斜，两面均无毛，侧脉每边 11~12 条，边全缘。花序腋上生，短于叶，长不及 30 cm。蒴果近卵形，长 2~2.5 cm，熟后橙黄色。花期 5~9 月，果期 10 月至翌年 4 月。

Trees, up to 20–30 m tall. Leaves odd-pinnate, leaflets 9–11(15), opposite, oblong, base oblique, cuneate, both furfaces glabrous, secondary veins 11–12 on each side, margin entire. Inflorescences axillary, shorter than leaves, less than 30 cm. Capsules nearly ovate, 2–2.5 cm long, orangish when mature. Fl. May–Sep., fr. Oct.–Apr. of next year.

嫩枝 New branch
摄影：郭屹立 Photo by：Guo Yili

复叶 Compound leaf
摄影：李冬兴 Photo by：Li Dongxing

果序和种子 Infructescence and seeds
摄影：黄健 Photo by：Huang Jian

• 1~5 cm DBH + 5~20 cm DBH ○ ≥20 cm DBH
个体分布图 Distribution of individuals

径级分布表 DBH class

胸径区间 (Diameter class) (cm)	个体数 (No. of individuals in the plot)	比例 (Proportion) (%)
1~2	1	8.33
2~5	4	33.33
5~10	4	33.33
10~20	2	16.67
20~35	1	8.33
35~60	0	0.00
≥60	0	0.00

144 麻楝 má liàn | Chittagong Chickrassy

Chukrasia tabularis A. Juss.
楝科 Meliaceae

代码（SpCode）= CHUTAB
个体数（Individual number/15 hm^2）= 57
最大胸径（Max DBH）= 16.3 cm
重要值排序（Importance value rank）= 103

乔木，高达 25 m。老茎树皮纵裂，无毛，具苍白色的皮孔。偶数羽状复叶，无毛，小叶 10~16 枚, 小叶互生，卵形，基部偏斜，侧脉每边 10~15 条，至边缘处分叉。圆锥花序顶生，长约为叶的一半。蒴果灰黄色或褐色，近球形或椭圆形，无毛。花期 4~5 月，果期 7 月至翌年 1 月。

Trees, up to 25 m tall. Bark of old branches exfoliating, glabrous, with pale lenticels. Leaves usually even-pinnate, glabrous, leaflets 10–16, alternate, papery, ovate, base oblique, secondary veins 10–15 on each side of midvein, bifurcate to the margin. Thrses terminal, ca. 1/2 as long as leaves. Capsules grawish yellow or brown, subglobose ro elliptic, glabrous. Fl. Apr.–May, fr. Jul.–Jan. of next year.

花枝 Flowering branches
摄影：黄俞淞 Photo by：Huang Yusong

花序 Inflorescence
摄影：黄俞淞 Photo by：Huang Yusong

复叶 Compound leaf
摄影：黄俞淞 Photo by：Huang Yusong

• 1~5 cm DBH + 5~20 cm DBH ○ ⩾20 cm DBH
个体分布图 Distribution of individuals

径级分布表 DBH class

胸径区间 (Diameter class) (cm)	个体数 (No. of individuals in the plot)	比例 (Proportion) (%)
1~2	21	36.84
2~5	23	40.35
5~10	6	10.53
10~20	7	12.28
20~35	0	0.00
35~60	0	0.00
⩾60	0	0.00

145 灰毛浆果楝

huī máo jiàng guǒ liàn | Grey-hair Cipadessa

Cipadessa cinerascens (Pell.)Hand.-Mazz.
楝科 Meliaceae

代码（SpCode）= CIPCIN
个体数（Individual number/15 hm^2）= 336
最大胸径（Max DBH）= 25.0 cm
重要值排序（Importance value rank）= 32

落叶灌木或小乔木，高 1~4 m，少数植株高达 8~10 m。树皮粗糙。小叶通常 4~6 对，对生，卵形至卵状长圆形，基部圆形或宽楔形，偏斜，被灰黄色柔毛。圆锥花序腋生，分枝伞房花序式。核果球形，熟后紫黑色。花期 4~10 月，果期 8~12 月。

Deciduous shrubs or small trees, 1–4 m tall, sometimes up to 8-10 m. Bark coarse. Leaflets usually 4–6 pairs, opposite, ovate to ovate-oblong, base rounded or broadly cuniform, oblique, covered with appressed yellowish gray pubescence. Thyrses axillary, branches corymbose. Drupes globose, purple to black when mature. Fl. Apr.–Oct., fr. Aug.–Dec..

树干 Trunk
摄影：王斌 Photo by: Wang Bin

花枝 Flowering branches
摄影：黄俞淞 Photo by: Huang Yusong

果序 Infructescence
摄影：刘晟源 Photo by: Liu Shengyuan

• 1~5 cm DBH + 5~20 cm DBH ○ ≥20 cm DBH
个体分布图 Distribution of individuals

径级分布表 DBH class

胸径区间 (Diameter class) (cm)	个体数 (No. of individuals in the plot)	比例 (Proportion) (%)
1~2	43	12.80
2~5	111	33.04
5~10	128	38.10
10~20	53	15.77
20~35	1	0.30
35~60	0	0.00
≥60	0	0.00

146 海南樫木

hǎi nán jiān mù | Very-soft Pencilwood

Dysoxylum mollissimum Blume
楝科 Meliaceae

代码（SpCode）= DYSMOL
个体数（Individual number/15 hm^2）= 2
最大胸径（Max DBH）= 12.2 cm
重要值排序（Importance value rank）= 186

乔木，通常高 7~10 m，有时超 20 m。小枝薄被柔毛。叶互生，羽状复叶，有小叶 20~30 枚，叶柄和叶轴被广展的长柔毛, 小叶对生或近对生，膜质，长圆形，叶面除中脉被毛外，其余无毛，背面被疏长毛。圆锥花序腋生，长约 18 cm，被疏柔毛。核果球形，熟后黄色。花期 5~9 月，果期 10~11 月。

Trees, usually 7–10 m tall, sometimes over 20 m. Leaves alternate, pinnate, leaflets 20–30, petiole and rachis villous, leaflets opposite or subopposite, membranous, oblong, adaxially glabrous except midvein, abaxially densely villous. Thyrses axillary, ca. 18 cm long, pubescence. Capsule globose, yellow when mature. Fl. May–Sep., fr. Oct.–Nov..

树干 Trunk
摄影：王斌 Photo by: Wang Bin

嫩枝叶 New branch and leaves
摄影：李冬兴 Photo by: Li Dongxing

复叶 Compound leaves
摄影：李冬兴 Photo by: Li Dongxing

• 1~5 cm DBH + 5~20 cm DBH ○ ≥20 cm DBH
个体分布图 Distribution of individuals

径级分布表 DBH class

胸径区间 (Diameter class) (cm)	个体数 (No. of individuals in the plot)	比例 (Proportion) (%)
1~2	0	0.00
2~5	0	0.00
5~10	1	50.00
10~20	1	50.00
20~35	0	0.00
35~60	0	0.00
≥60	0	0.00

147 香椿

xiāng chūn | Chinese Toona

Toona sinensis (A. Juss.) Roem.
楝科 Meliaceae

代码（SpCode）= TOOSIN
个体数（Individual number/15 hm^2）= 1
最大胸径（Max DBH）= 8.9 cm
重要值排序（Importance value rank）= 207

乔木。树皮粗糙。偶数羽状复叶，小叶16~20，对生或互生，纸质，卵状披针形或卵状长椭圆形，不对称，边全缘或有疏离的小锯齿，两面均无毛，侧脉每边18~24条。圆锥花序与叶等长或更长，被稀疏的锈色短柔毛。核果狭椭圆形，长2~3.5厘米。花期6~8月，果期10~12月。

Trees. Bark rough. Leaves even-pinnate, leaflets 16–20, opposite or alternate, papery, ovate-lanceolate or ovate-elliptic, base asymmetric. margin entire or with alienated serrate, both surfaces glabrous, secondary veins 18–24 on each side of midvein. Panicle as long as leaves, or much longer, with short appressed. Capsule narrowly elliptic, 2–3.5 cm long. Fl. Jun.–Aug., fr. Oct.–Dec..

植株 Whole plant
摄影：黄俞淞 Photo by：Huang Yusong

果枝 Fruiting branches
摄影：黄俞淞 Photo by：Huang Yusong

果序 Infructescence
摄影：黄俞淞 Photo by：Huang Yusong

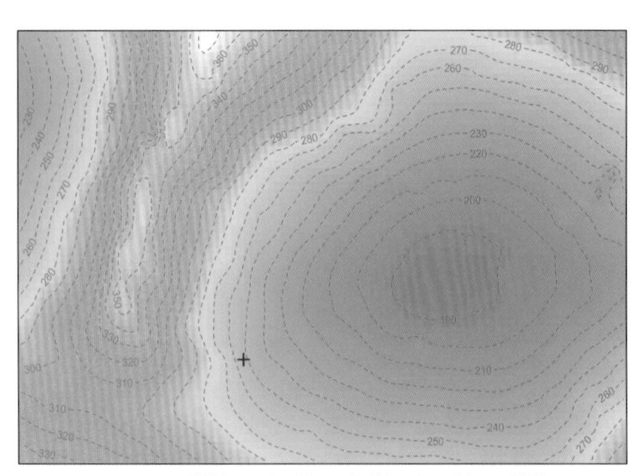
• 1~5 cm DBH + 5~20 cm DBH ○ ≥20 cm DBH
个体分布图 Distribution of individuals

径级分布表 DBH class

胸径区间 (Diameter class) (cm)	个体数 (No. of individuals in the plot)	比例 (Proportion) (%)
1~2	0	0.00
2~5	0	0.00
5~10	1	100.00
10~20	0	0.00
20~35	0	0.00
35~60	0	0.00
≥60	0	0.00

148 割舌树

gē shé shù | Robust Walsura

Walsura robusta Roxb.
楝科 Meliaceae

代码（SpCode）= WALROB
个体数（Individual number/15 hm^2）= 1315
最大胸径（Max DBH）= 39.4 cm
重要值排序（Importance value rank）= 20

常绿乔木，高 10~25 m。枝褐色。有小叶 3~5 片，小叶对生，长椭圆形或披针形，小叶柄两端膨大，具节。圆锥花序，疏被粉状短柔毛。浆果球形或卵形，密被黄褐色柔毛。花期 2~3 月，果期 4~6 月。

Evergreen trees, 10–25 m tall. Branches brown. Leaflets 3–5, opposite, oblong-elliptic or lanceolate, ends of petiole swollen, jointed. Panicle sparsely pubescent. Berry globose or ovate, densely covered with yellowish gray villous. Fl. Feb.–Mar., fr. Apr.–Jun..

树干 Trunk
摄影：王斌 Photo by：Wang Bin

枝叶 Branch and leaves
摄影：黄俞淞 Photo by：Huang Yusong

果 Fruit
摄影：黄俞淞 Photo by：Huang Yusong

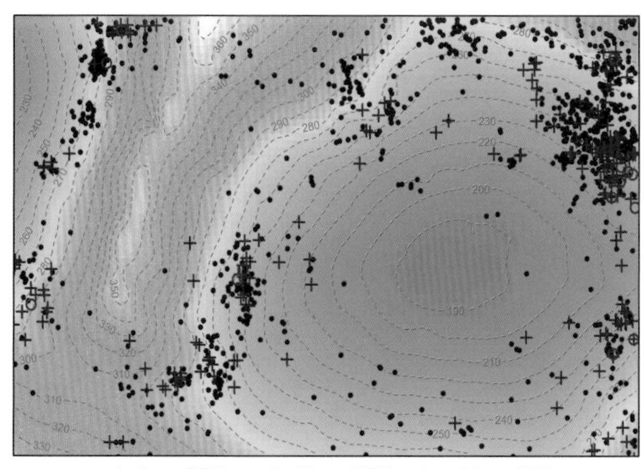

• 1~5 cm DBH + 5~20 cm DBH ○ ⩾20 cm DBH
个体分布图 Distribution of individuals

径级分布表 DBH class

胸径区间 (Diameter class) (cm)	个体数 (No. of individuals in the plot)	比例 (Proportion) (%)
1~2	670	50.95
2~5	429	32.62
5~10	138	10.49
10~20	64	4.87
20~35	12	0.91
35~60	2	0.15
⩾60	0	0.00

149 波叶异木患

bō yè yì mù huàn | Caudata Allophylus

Allophylus caudatus Radlk.
无患子科 Sapindaceae

代码（SpCode）= ALLCAU
个体数（Individual number/15 hm^2）= 30
最大胸径（Max DBH）= 2.1 cm
重要值排序（Importance value rank）= 125

灌木或小乔木，通常不高于 5 m。叶具三小叶，小叶片膜质或薄纸质，叶背无毛，叶面中脉具毛，侧脉约 10 对，边缘具波浪状齿。总状花序，不分枝。果红色，近球形，直径 7~8 mm。花期 8~9 月，果期 9~11 月。

Shrubs or small trees, usually less than 5 m tall. Leaves trifoliate, leaflet blade membranous or thinly papery, abaxially glabrous, adaxially hairy on midvein, lateral veins ca. 10 pairs, margin shallowly wavy toothed. Inflorescences racemose, main rachis unbranched. Fruit red, subglobose, 7–8 mm in diam.. Fl. Aug.–Sep., fr. Sep.–Nov..

枝叶 Branch and leaves
摄影：黄俞淞 Photo by：Huang Yusong

果 Fruits
摄影：丁涛 Photo by：Ding Tao

复叶 Compound leaf
摄影：黄俞淞 Photo by：Huang Yusong

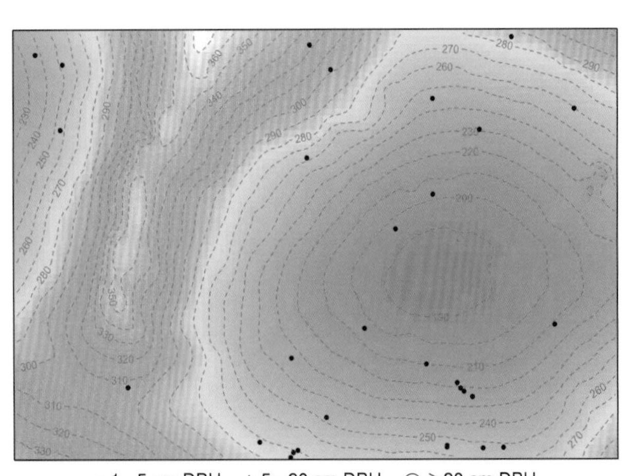

• 1~5 cm DBH + 5~20 cm DBH ○ ≥20 cm DBH

个体分布图 Distribution of individuals

径级分布表 DBH class

胸径区间 (Diameter class) (cm)	个体数 (No. of individuals in the plot)	比例 (Proportion) (%)
1~2	28	93.33
2~5	2	6.67
5~10	0	0.00
10~20	0	0.00
20~35	0	0.00
35~60	0	0.00
≥60	0	0.00

150 细子龙

xì zǐ lóng | Chinese Amesiodendron

Amesiodendron chinense (Merr.) Hu
无患子科 Sapindaceae

代码（SpCode）= AMECHI
个体数（Individual number/15 hm^2）= 59
最大胸径（Max DBH）= 25.5 cm
重要值排序（Importance value rank）= 108

常绿乔木，高可达 25 m。树皮暗灰色。小叶 3~7 对，薄革质，基端一对小叶卵形，其余的长圆形或长圆状披针形，两侧稍不对称，边缘皱波状。花序常几个丛生于小枝的顶端，密被短柔毛。蒴果的发育果爿近球形，外面有瘤状凸起和密集的淡褐色皮孔。花期 5 月，果期 8~10 月。

Evergreen trees, up to 25 m tall. Bark dark gray. Leaflets 3–7 pairs, thinly leathery, first pair ovate, others oblong or oblong-lanceolate, bilaterally slightly asymmetrical, margin rugose wavy. Inflorescences often several fascicled at branch apex, densely tomentose. Schizocarps subglobose, slightly to coarsely tuberculous and with dense, pale brown small lenticels. Fl. May, fr. Aug.–Oct..

树干 Trunk
摄影：王斌 Photo by：Wang Bin

枝叶 Branch and leaves
摄影：黄俞淞 Photo by：Huang Yusong

叶背 Leaf back
摄影：王斌 Photo by：Wang Bin

• 1~5 cm DBH + 5~20 cm DBH ○ ⩾20 cm DBH
个体分布图 Distribution of individuals

径级分布表 DBH class

胸径区间 (Diameter class) (cm)	个体数 (No. of individuals in the plot)	比例 (Proportion) (%)
1~2	19	32.20
2~5	22	37.29
5~10	8	13.56
10~20	7	11.86
20~35	3	5.08
35~60	0	0.00
⩾60	0	0.00

151 黄梨木

huáng lí mù | Small Boniodendron

Boniodendron minus (Hemsl.) T. C. Chen
无患子科 Sapindaceae

代码（SpCode）= BONMIN
个体数（Individual number/15 hm^2）= 694
最大胸径（Max DBH）= 35.8 cm
重要值排序（Importance value rank）= 15

落叶乔木，高可达 25 m。树皮暗褐色，具纵裂纹。叶聚生于小枝先端，一回偶数羽状复叶，小叶 10~20 片，披针形或椭圆形。聚伞圆锥花序顶生，稍有腋生，被短柔毛，花淡黄色至白色。蒴果近球形，具 3 翅，顶端凹入且具宿存花柱。花期 5~6 月，果期 7~8 月。

Deciduous trees, up to 25 m tall. Bark dark brown, fissured. Leaves fascicled at branch apices, paripinnate, leaflets 10–20, lanceolate or elliptic. Thyrses terminal, rarely axillary, pubescent, flowers yellowish to white. Capsules subglobose, 3-winged, apex concave and with persistent style. Fl. May–Jun., fr. Jul.–Aug..

树干 Trunk
摄影：王斌 Photo by：Wang Bin

果序 Infructescence
摄影：黄俞淞 Photo by：Huang Yusong

复叶 Compound leaves
摄影：丁涛 Photo by：Ding Tao

• 1~5 cm DBH + 5~20 cm DBH ○ ≥20 cm DBH
个体分布图 Distribution of individuals

径级分布表 DBH class

胸径区间 (Diameter class) (cm)	个体数 (No. of individuals in the plot)	比例 (Proportion) (%)
1~2	111	15.99
2~5	168	24.21
5~10	152	21.90
10~20	205	29.54
20~35	56	8.07
35~60	2	0.29
≥60	0	0.00

152 茶条木

chá tiáo mù | Yunnan Delavaya

Delavaya toxocarpa Franch.
无患子科 Sapindaceae

代码（SpCode）= DELTOX
个体数（Individual number/15 hm^2）= 70
最大胸径（Max DBH）= 13.1 cm
重要值排序（Importance value rank）= 76

落叶灌木或小乔木，高 3~8 m。树皮褐红色。叶薄革质，全部小叶边缘均有粗锯齿，少全缘。花序狭窄，柔弱而疏花。蒴果深紫色，裂片长 1.5~2.5 cm。花期 4 月，果期 8 月。

Deciduous shrubs or small trees, 3–8 m tall. Bark brownish red. Leaf blade thinly leathery, margin somewhat thickly serrate, rarely entire. Inflorescences narrow, slender, aparsely flowered. Capsules royal purple, lobes 1.5–2.5 cm. Fl. Apr., fr. Aug..

树干 Trunk
摄影：王斌 Photo by: Wang Bin

果枝 Fruiting branch
摄影：黄俞淞 Photo by: Huang Yusong

花 Flower
摄影：黄俞淞 Photo by: Huang Yusong

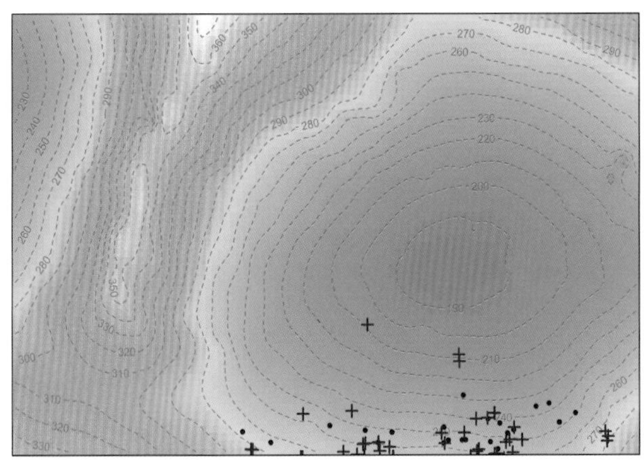

• 1～5 cm DBH + 5～20 cm DBH ○ ≥20 cm DBH
个体分布图 Distribution of individuals

径级分布表 DBH class

胸径区间 (Diameter class) (cm)	个体数 (No. of individuals in the plot)	比例 (Proportion) (%)
1~2	15	21.43
2~5	12	17.14
5~10	31	44.29
10~20	12	17.14
20~35	0	0.00
35~60	0	0.00
≥60	0	0.00

153 茎花赤才

jìng huā chì cái | Cauliflorous Lepisanthes

Lepisanthes cauliflora C. F. Liang et S. L. Mo
无患子科 Sapindaceae

代码（SpCode）= LEPCAU
个体数（Individual number/15 hm^2）= 85
最大胸径（Max DBH）= 3.2 cm
重要值排序（Importance value rank）= 110

常绿灌木或小乔木，高 2~6 m。偶数羽状复叶，互生，小叶 4-6 对，披针形，对生或近对生，叶背被毛，叶脉在背面明显。总状花序或聚伞花序，茎生或老枝上。果球形或扁球形，具 3 钝棱。花期 9~10 月。

Evergreen shrubs or small trees, 2–6 m tall. Leaves even-pinnately, alternate, leaflets 4–6 pairs, lanceolate, opposite or subopposite, abaxially pubescent, veins prominent. Inflorescences racemose or thyrsoid, on stems and old branches. Fruit globous or compressed globose, with 3 obtuse angles. Fl. Sep.–Oct..

果 Fruits
摄影：黄俞淞 Photo by: Huang Yusong

花序 Inflorescence
摄影：黄俞淞 Photo by: Huang Yusong

复叶 Compound leaves
摄影：黄俞淞 Photo by: Huang Yusong

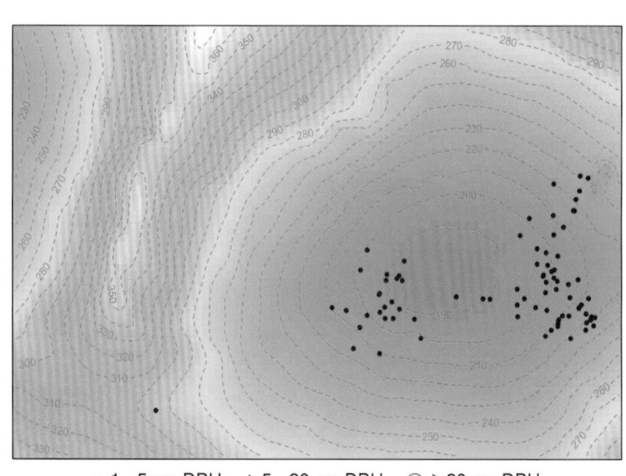

• 1~5 cm DBH + 5~20 cm DBH ○ ≥20 cm DBH
个体分布图 Distribution of individuals

径级分布表 DBH class

胸径区间 (Diameter class) (cm)	个体数 (No. of individuals in the plot)	比例 (Proportion) (%)
1~2	70	82.35
2~5	15	17.65
5~10	0	0.00
10~20	0	0.00
20~35	0	0.00
35~60	0	0.00
≥60	0	0.00

154 弄岗滇赤才

nòng gǎng diān chì cái | Longgang Aphania

Aphania longgangensis Y.S.Huang
无患子科 Sapindaceae

代码（SpCode）= APHLON
个体数（Individual number/15 hm^2）= 62
最大胸径（Max DBH）= 21.9 cm
重要值排序（Importance value rank）= 113

灌木或小乔木，高 4~8 m。小枝圆柱形，灰白色，无毛。偶数羽状复叶，互生，无托叶，小叶通常 1~2 对，对生或近对生，椭圆状披针形，基部楔形。花果期未见。

Shrubs or small trees, 4–8 m tall. Branchlets terete, gray white, glabrous. Leaves paripinnate, alternate, without stipule, leaflets 1–2 pairs, opposite or subopposite, elliptic-lanceolate, base cuneate. Fl. and fr. not seen.

树干 Trunk
摄影：王斌 Photo by：Wang Bin

叶 Leaves
摄影：王斌 Photo by：Wang Bin

叶背 Leaf back
摄影：王斌 Photo by：Wang Bin

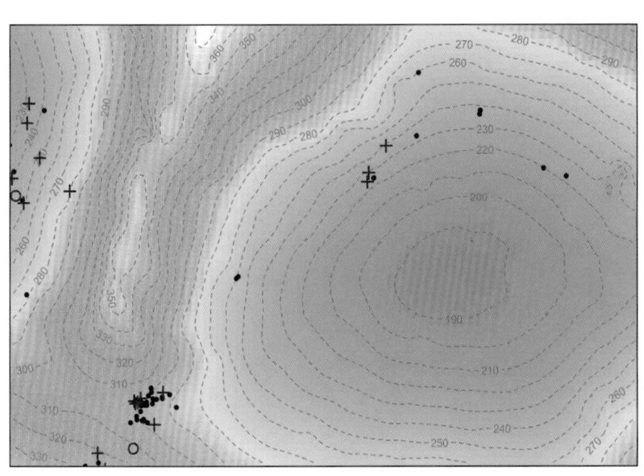

• 1~5 cm DBH + 5~20 cm DBH ○ ≥20 cm DBH
个体分布图 Distribution of individuals

径级分布表 DBH class

胸径区间 (Diameter class) (cm)	个体数 (No. of individuals in the plot)	比例 (Proportion) (%)
1~2	18	29.03
2~5	25	40.32
5~10	12	19.35
10~20	5	8.06
20~35	2	3.23
35~60	0	0.00
≥60	0	0.00

155 粗柄槭

cū bǐng qī | Tonkin Maple

Acer tonkinense Lecomte
槭树科 Aceraceae

代码（SpCode）= ACETON
个体数（Individual number/15 hm^2）= 82
最大胸径（Max DBH）= 20.1 cm
重要值排序（Importance value rank）= 107

落叶乔木，高 8~12 m。树皮深褐色，有蜡质白粉。叶近革质，基部近于圆形或心脏形，中部以上 3 裂，主脉 3 条。花序圆锥状，顶生。小坚果近卵圆形，嫩时淡紫色，成熟后淡黄色。花期 4~5 月，果期 9 月。

Deciduous trees, 8–12 m tall. Bark dark brown, branchlets white waxy. Leaf blade subleathery, base nearly rounded or cordate, deeply or shallowly 3-lobed, primary veins 3. Inflorescence terminal, paniculate. Nutlets subovate, purplish when young, yellowish when mature. Fl. Apr.–May, fr. Sep..

果枝 Fruiting branches
摄影：黄俞淞 Photo by: Huang Yusong

叶背 Leaf back
摄影：丁涛 Photo by: Ding Tao

果 Fruits
摄影：黄俞淞 Photo by: Huang Yusong

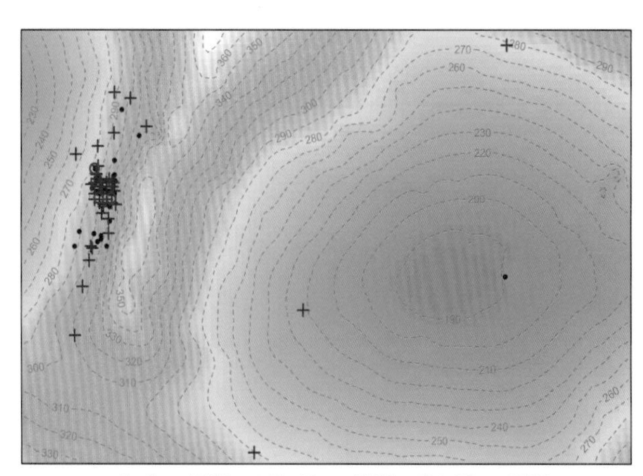
• 1~5 cm DBH + 5~20 cm DBH ○ ≥20 cm DBH
个体分布图 Distribution of individuals

径级分布表 DBH class

胸径区间 (Diameter class) (cm)	个体数 (No. of individuals in the plot)	比例 (Proportion) (%)
1~2	5	6.10
2~5	30	36.59
5~10	26	31.71
10~20	18	21.95
20~35	3	3.66
35~60	0	0.00
≥60	0	0.00

156 山様叶泡花树

shān shēn yè pào huā shù | Thorel's Meliosma

Meliosma thorelii Lecomte
清风藤科 Sabiaceae

代码（SpCode）＝ MELTHO
个体数（Individual number/15 hm^2）＝ 4
最大胸径（Max DBH）＝ 4.1 cm
重要值排序（Importance value rank）＝ 178

常绿乔木，高 6~14 m。小枝圆柱形。单叶，革质、倒披针状椭圆形或倒披针形，全缘或中上部有锐尖的小锯齿，脉腋有髯毛。圆锥花序顶生或生于上部叶腋，直立。核果球形，直径 6~9 mm。花期夏季，果期 10~11 月。

Evergreen trees, 6–14 m tall. Branchlets terete. Leaves simple, leathery, oblanceolate-elliptic or oblanceolate, margin entire or acutely sarrulate, vein axils crinite. Panicle terminal or axillary on apical branches, erect. Drupe globose, 6–9 mm in diam. Fl. summer, fr. Oct.–Nov..

树干 Trunk
摄影：黄俞淞 Photo by：Huang Yusong

花序 Inflorescence
摄影：黄俞淞 Photo by：Huang Yusong

枝叶 Branch and leaves
摄影：黄俞淞 Photo by：Huang Yusong

• 1~5 cm DBH　+ 5~20 cm DBH　○ ≥20 cm DBH
个体分布图 Distribution of individuals

径级分布表 DBH class

胸径区间 (Diameter class) (cm)	个体数 (No. of individuals in the plot)	比例 (Proportion) (%)
1~2	1	25.00
2~5	3	75.00
5~10	0	0.00
10~20	0	0.00
20~35	0	0.00
35~60	0	0.00
≥60	0	0.00

157 南酸枣

nán suān zǎo | Nepali Hog-plum

Choerospondias axillaris (Roxb.) Burtt et Hill
漆树科 Anacardiaceae

代码（SpCode）= CHOAXI
个体数（Individual number/15 hm^2）= 41
最大胸径（Max DBH）= 71.7 cm
重要值排序（Importance value rank）= 72

落叶乔木，高 8~20 m。树皮灰褐色，片状剥落。奇数羽状复叶长 25~40 cm，有小叶 3~6 对，叶轴无毛，卵形或卵状披针形，基部偏斜。雄花序长 4~10 cm，被微柔毛，雌花单生于上部叶腋。核果倒卵状椭圆形，成熟时黄色，果核顶端具 5 个小孔。花期 4 月，果期 8~10 月。

Deciduous trees, 8–20 m tall. Bark gray brown, exfoliate. Odd-pinnately, leaves 25–40 cm long, with 3–6 pairs leaflets, rachis glabrous, leaflets membranous to papery, ovate-lanceolate, base oblique. Male inflorescence 4–10 cm long, pubescent to glabrous, female flowers solitary in axils of distal leaves. Drupe ovate-elliptic, yellow when mature, 5-locular of distal nutlet. Fl. Apr., fr. Aug.–Oct..

树干 Trunk
摄影：王斌 Photo by：Wang Bin

果 Fruits
摄影：刘晟源 Photo by：Liu Shengyuan

果枝 Fruiting branches
摄影：黄俞淞 Photo by：Huang Yusong

径级分布表 DBH class

胸径区间 (Diameter class) (cm)	个体数 (No. of individuals in the plot)	比例 (Proportion) (%)
1~2	6	14.63
2~5	7	17.07
5~10	6	14.63
10~20	6	14.63
20~35	10	24.39
35~60	5	12.20
⩾60	1	2.44

• 1~5 cm DBH + 5~20 cm DBH ○ ⩾20 cm DBH
个体分布图 Distribution of individuals

158 黄连木

huáng lián mù | Chinese Pistache

Pistacia chinensis Bunge
漆树科 Anacardiaceae

代码（SpCode）= PISCHI
个体数（Individual number/15 hm^2）= 35
最大胸径（Max DBH）= 29.4 cm
重要值排序（Importance value rank）= 104

落叶乔木，高达 20 m。树皮暗褐色，呈鳞片状剥落。奇数羽状复叶互生，有小叶 5~6 对，对生或近对生，纸质，披针形或卵状披针形或线状披针形，基部偏斜，全缘。圆锥花序腋生，先花后叶。核果倒卵状球形，直径约 5 mm，成熟时紫红色。花期 3~5 月，果期 8~11 月。

Deciduous trees, up to 20 m tall. Bark dark brown, exfoliate. Odd-pinnate, alternate, leaflets 5–6 pairs, opposite to subopposite, papery, lanceolate, ovate-lanceolate or linear-lanceolate, base oblique, margin entire. Panicle axillary, flowers produced before leafing. Drupe obovate-globose, ca. 5 mm in diam., carmine when mature. Fl. Mar.–May, fr. Aug.–Nov..

树干 Trunk
摄影：黄俞淞 Photo by：Huang Yusong

枝叶 Branch and leaves
摄影：黄俞淞 Photo by：Huang Yusong

复叶 Compound leaf
摄影：黄俞淞 Photo by：Huang Yusong

• 1~5 cm DBH + 5~20 cm DBH ○ ≥20 cm DBH
个体分布图 Distribution of individuals

径级分布表 DBH class

胸径区间 (Diameter class) (cm)	个体数 (No. of individuals in the plot)	比例 (Proportion) (%)
1~2	1	2.86
2~5	4	11.43
5~10	10	28.57
10~20	17	48.57
20~35	3	8.57
35~60	0	0.00
≥60	0	0.00

159 清香木

qīng xiāng mù | Yunnan Pistache

Pistacia weinmanniifolia J. Poiss. ex Franch.
漆树科 Anacardiaceae

代码（SpCode）= PISWEI
个体数（Individual number/15 hm^2）= 225
最大胸径（Max DBH）= 25.1 cm
重要值排序（Importance value rank）= 55

落叶灌木或小乔木，高 2~15 m。树皮灰色。偶数羽状复叶，小叶革质，长圆形或倒卵状长圆形，先端圆或微缺，具芒刺状硬尖头。花序腋生，花无柄。核果球形，成熟时红色。花期 3~5 月，果期 6~8 月。

Deciduous shrubs or small trees, 2–15 m tall. Bark gray. Even-pinnately, leaflet blade leathery, oblong or obovate-oblong, apex rounded or emarginated, mucronate. Inflorescence axillary, flowers sessile. Drupe globose, red when mature. Fl. Mar.–May, fr. Jun.–Aug..

树干 Trunk
摄影：王斌 Photo by：Wang Bin

果枝 Fruiting branches
摄影：黄俞淞 Photo by：Huang Yusong

复叶 Compound leaf
摄影：王斌 Photo by：Wang Bin

• 1~5 cm DBH + 5~20 cm DBH ○ ≥20 cm DBH
个体分布图 Distribution of individuals

径级分布表 DBH class

胸径区间 (Diameter class) (cm)	个体数 (No. of individuals in the plot)	比例 (Proportion) (%)
1~2	33	14.67
2~5	68	30.22
5~10	64	28.44
10~20	56	24.89
20~35	4	1.78
35~60	0	0.00
≥60	0	0.00

160 大叶清香木 dà yè qīng xiāng mù | Bigflower Pistache

Pistacia species No.1
漆树科 Anacardiaceae

代码（SpCode）= PISSP1
个体数（Individual number/15 hm^2）= 2
最大胸径（Max DBH）= 11.9 cm
重要值排序（Importance value rank）= 188

常绿灌木或小乔木，高 2~6 m。树皮灰色。偶数羽状复叶，小叶对生或近对生，椭圆状披针形，长 4~7 cm，宽 1.5~2.5 cm，先端通常具小尖头，基部不对称，叶轴上面具槽。花序腋生。核果扁球形，长约 7 mm，径约 6 mm，成熟时红色。花期 5~7 月，果期 8~10 月。

Evergreen shrubs or small trees, 2–6 m tall. Bark gray. Leaves paripinnate, leaflets opposite or subopposite, elliptic-lanceolate, 4–7 cm long, and 1.5–2.5 cm wide, apex usually mucronate，base oblique, leaf rachis grooved. Inflorescence axillary. Drupe subglobose, ca. 7 mm long, and ca. 6 mm in diam., red when mature. Fl. May–Jul., fr. Aug.–Oct..

果枝 Fruiting branches
摄影：黄俞淞 Photo by：Huang Yusong

果序 Infructescence
摄影：黄俞淞 Photo by：Huang Yusong

复叶 Compound leaves
摄影：黄俞淞 Photo by：Huang Yusong

• 1~5 cm DBH + 5~20 cm DBH ○ ⩾20 cm DBH
个体分布图 Distribution of individuals

径级分布表 DBH class

胸径区间 (Diameter class) (cm)	个体数 (No. of individuals in the plot)	比例 (Proportion) (%)
1~2	0	0.00
2~5	1	50.00
5~10	0	0.00
10~20	1	50.00
20~35	0	0.00
35~60	0	0.00
⩾60	0	0.00

161 滨盐肤木

bīn yán fú mù | Roxburgh's Sumac

Rhus chinensis var. *roxburghii* (DC.) Rehd.
漆树科 Anacardiaceae

代码（SpCode）= RHUCHI
个体数（Individual number/15 hm^2）= 2
最大胸径（Max DBH）= 4.0 cm
重要值排序（Importance value rank）= 196

落叶小乔木或灌木，高 2~10 m。小枝棕褐色，被锈色柔毛，具圆形小皮孔。奇数羽状复叶有小叶约 6 对，叶轴无叶状翅，小叶椭圆状卵形，叶背粉绿色，被白粉。圆锥花序多分枝，雄花序长 30~40 cm，雌花序较短，密被锈色柔毛。核果球形，成熟时红色。花期 8~9 月，果期 10 月。

Deciduous small trees or shrubs, 2–10 m tall. Branchlets brown, rust colored pubescent, lenticellate. Odd-pinnately, leaflets ca. 6 pairs, rachis wingless, leaflets ovate , elliptic-ovate or oblong, abaxially glaucous. Panicle many branched, male inflorescence 30–40 cm long, female inflorescence short, densely rust colored pubescent. Drupes globose, red when mature. Fl. Aug.–Sep., fr. Oct..

枝叶 Branch and leaves
摄影：刘晟源 Photo by：Liu Shengyuan

果枝 Fruiting branch
摄影：黄俞淞 Photo by：Huang Yusong

果序 Infructescence
摄影：黄俞淞 Photo by：Huang Yusong

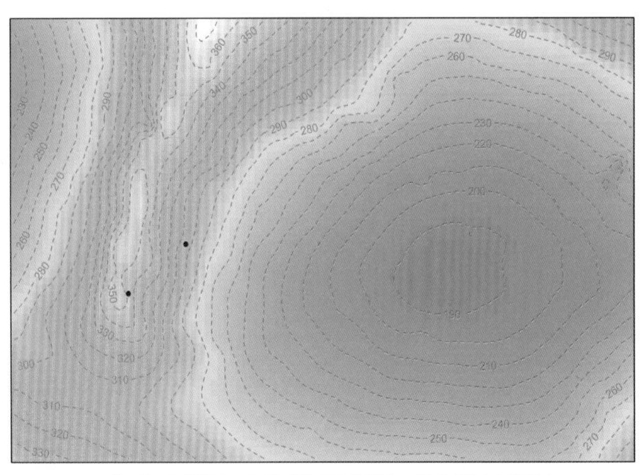

• 1~5 cm DBH　+ 5~20 cm DBH　○ ≥20 cm DBH

个体分布图 Distribution of individuals

径级分布表 DBH class

胸径区间 (Diameter class) (cm)	个体数 (No. of individuals in the plot)	比例 (Proportion) (%)
1~2	0	0.00
2~5	2	100.00
5~10	0	0.00
10~20	0	0.00
20~35	0	0.00
35~60	0	0.00
≥60	0	0.00

162 岭南酸枣

lǐng nán suān zǎo | Canton Mombin

Spondias lakonensis Pierre
漆树科 Anacardiaceae

代码（SpCode）= SPOLAK
个体数（Individual number/15 hm^2）= 16
最大胸径（Max DBH）= 56.9 cm
重要值排序（Importance value rank）= 95

落叶乔木，高 8~15 m。小枝灰褐色，疏被微柔毛。叶互生，奇数羽状复叶长 25~35 cm，有小叶 5~11 对，叶轴和叶柄圆柱形，疏被微柔毛，小叶长圆状披针形，全缘。圆锥花序腋生。核果倒卵状，成熟时带红色，果核顶端具 4 角和 4 个凹点。花期 3~4 月，果期 5~9 月。

Deciduous trees, 8–15 m tall. Branchlets grayish brown, sparsely pubescent. Leaves alternate, odd-pinnately, 25–35 cm long, leaflets 5–11 pairs, rachis and petiole terete, sparsely pubescent, leaflet oblong-lanceolate, margin entire. Panicle axillary. Drupes obovate, red when mature, apex of nutlet with 4-angled and 4-locular. Fl. Mar.–Apr., fr. May–Sep..

植株 Whole plant
摄影：黄俞淞 Photo by：Huang Yusong

复叶 Compound leaves
摄影：黄俞淞 Photo by：Huang Yusong

果序 Infructescence
摄影：黄俞淞 Photo by：Huang Yusong

• 1~5 cm DBH + 5~20 cm DBH ○ ≥20 cm DBH
个体分布图 Distribution of individuals

径级分布表 DBH class

胸径区间 (Diameter class) (cm)	个体数 (No. of individuals in the plot)	比例 (Proportion) (%)
1~2	0	0.00
2~5	1	6.25
5~10	2	12.50
10~20	1	6.25
20~35	8	50.00
35~60	4	25.00
≥60	0	0.00

163 野漆

yě qī | Wax Tree

Toxicodendron succedaneum (L.) Kuntze
漆树科 Anacardiaceae

代码（SpCode）= TOXSUC
个体数（Individual number/15 hm^2）= 117
最大胸径（Max DBH）= 33.8 cm
重要值排序（Importance value rank）= 65

落叶乔木或小乔木，高达 10 m。奇数羽状复叶互生，常集生小枝顶端，小叶对生或近对生，坚纸质至薄革质。圆锥花序多分枝，花黄绿色。核果大，偏斜，压扁。花期 5~6 月，果期 7~10 月。

Deciduous trees or small trees, up to 10 m. Odd-pinnately, leaves alternate, usually clustered branchlets apex, leaflets opposite or subopposite, thickly papery to thinly leathery. Panicle many branched, flowers yellowish green. Drupes asymmetrical, compressed. Fl. May–Jun., fr. Jul.–Oct..

果枝 Fruiting branch
摄影：黄俞淞 Photo by：Huang Yusong

果序 Infructescence
摄影：黄俞淞 Photo by：Huang Yusong

复叶 Compound leaf
摄影：黄俞淞 Photo by：Huang Yusong

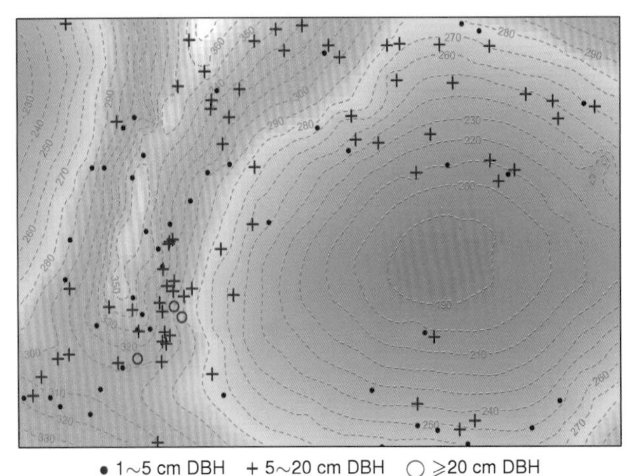

• 1~5 cm DBH + 5~20 cm DBH ○ ≥20 cm DBH
个体分布图 Distribution of individuals

径级分布表 DBH class

胸径区间 (Diameter class) (cm)	个体数 (No. of individuals in the plot)	比例 (Proportion) (%)
1~2	21	17.95
2~5	24	20.51
5~10	37	31.62
10~20	32	27.35
20~35	3	2.56
35~60	0	0.00
≥60	0	0.00

164 八角枫

bā jiǎo fēng | Chinese Alangium

Alangium chinense (Lour.) Harms subsp.*chinense*
八角枫科 Alangiaceae

代码（SpCode）= ALACHI
个体数（Individual number/15 hm^2）= 23
最大胸径（Max DBH）= 23.1 cm
重要值排序（Importance value rank）= 141

落叶乔木或灌木，高 3~15 m。叶纸质，近椭圆形、卵形，基部两侧常不对称，叶上面无毛，下面脉腋有丛状毛；基出脉 3~5(7)，成掌状。聚伞花序腋生，长 3~4 cm，被稀疏微柔毛。核果卵圆形，成熟后黑色。花期 5~7 月和 9~10 月，果期 7~11 月。

Deciduous trees or shrubs, 3–15 m tall. Leaves papery, elliptic or ovate, base asymmetried, adaxially glabrous, tufted pubescent at axils of veins, base veins 3–5(7), palmate. Panicle axillary, 3–4 cm long, sparsely pubescent. Drupes ovate-rounded, black when mature. Fl. May–Jul. and Sep.–Oct., fr. Jul.–Nov..

花枝 Flowering branch
摄影：丁涛 Photo by：Ding Tao

花序 Inflorescence
摄影：黄俞淞 Photo by：Huang Yusong

果枝 Fruiting branch
摄影：黄俞淞 Photo by：Huang Yusong

• 1~5 cm DBH + 5~20 cm DBH ○ ≥20 cm DBH
个体分布图 Distribution of individuals

径级分布表 DBH class

胸径区间 (Diameter class) (cm)	个体数 (No. of individuals in the plot)	比例 (Proportion) (%)
1~2	7	30.43
2~5	12	52.17
5~10	2	8.70
10~20	1	4.35
20~35	1	4.35
35~60	0	0.00
≥60	0	0.00

165 云山八角枫

yún shān bā jiǎo fēng | Handel's Alangium

Alangium kurzii var. *handelii* (Schnarf) Fang
八角枫科 Alangiaceae

代码（SpCode）= ALAKUR
个体数（Individual number/15 hm^2）= 54
最大胸径（Max DBH）= 19.2 cm
重要值排序（Importance value rank）= 111

灌木或小乔木，高 3~5 m。叶矩圆状卵形，稀椭圆形或卵形，长 11~19 cm，幼时两面有毛，其后无毛，叶柄长 2~2.5 cm。聚伞花序长 2.5~4 cm，花丝长 4 mm，有粗伏毛。核果椭圆形，长 8~10 mm。花期 5 月，果期 8 月。

Shrubs or small trees, 3–5 m tall. Leaves suborbicular, rarely rounded or ovate, 11–19 cm long, both surfaces pubescent when young, glabrous when old, petiole 2–2.5 cm. Cymes 2.5–4 cm long, appressed hirsute. Drupes elliptic, 8–10 mm long. Fl. May, fr. Aug..

叶 Leaves
摄影：黄俞淞 Photo by：Huang Yusong

果 Fruits
摄影：黄俞淞 Photo by：Huang Yusong

果枝 Fruiting branch
摄影：黄俞淞 Photo by：Huang Yusong

• 1~5 cm DBH + 5~20 cm DBH ○ ≥20 cm DBH
个体分布图 Distribution of individuals

径级分布表 DBH class

胸径区间 (Diameter class) (cm)	个体数 (No. of individuals in the plot)	比例 (Proportion) (%)
1~2	2	3.70
2~5	16	29.63
5~10	18	33.33
10~20	18	33.33
20~35	0	0.00
35~60	0	0.00
≥60	0	0.00

166 白花鹅掌柴 bái huā é zhǎng chái | White-flower Schefflera

Schefflera leucantha R. Vig.
五加科 Araliaceae

代码（SpCode）= SCHLEU
个体数（Individual number/15 hm^2）= 13
最大胸径（Max DBH）= 8.7 cm
重要值排序（Importance value rank）= 142

常绿灌木或攀缘灌木，高 2 m。小枝干时有纵皱纹。掌状复叶具小叶 5~7 枚，革质，长圆状披针形或椭圆状长圆形，边缘全缘，反卷。圆锥花序顶生；果实卵形，有 5 棱，黄红色。花期 4 月，果期 5 月。

Evergreen shrubs or climbers, 2 m tall. Branchlet longitudinal rugate when dry. Leaflets 5–7, leathery, oblong-lanceolate or elliptic-oblong, margin entire, deflexed. Panicle terminal, fruit ovate, 5-ribbed, orange-red. Fl. Apr., fr. May.

复叶 Compound leaf
摄影：黄俞淞 Photo by：Huang Yusong

花序 Inflorescence
摄影：黄俞淞 Photo by：Huang Yusong

果序 Infructescence
摄影：黄俞淞 Photo by：Huang Yusong

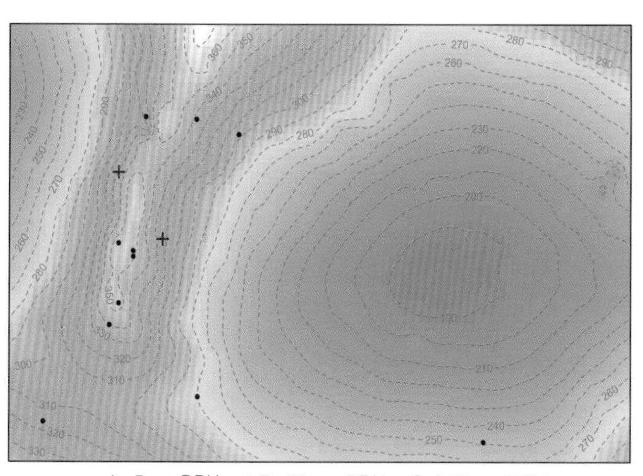

• 1~5 cm DBH + 5~20 cm DBH ○ ≥20 cm DBH
个体分布图 Distribution of individuals

径级分布表 DBH class

胸径区间 (Diameter class) (cm)	个体数 (No. of individuals in the plot)	比例 (Proportion) (%)
1~2	5	38.46
2~5	6	46.15
5~10	2	15.38
10~20	0	0.00
20~35	0	0.00
35~60	0	0.00
≥60	0	0.00

167 谅山鹅掌柴

liàng shān é zhǎng chái | Loc's Schefflera

Schefflera lociana Grushv. et Skvortsova
五加科 Araliaceae

代码（SpCode）= SCHLOC
个体数（Individual number/15 hm^2）= 2
最大胸径（Max DBH）= 3.7 cm
重要值排序（Importance value rank）= 198

常绿灌木或小乔木。叶为掌状复叶，小叶片 5~7 枚，薄革质，较厚大，叶背叶脉疏被星状柔毛，边缘全缘，反卷，叶柄可长达 1 m。果卵形，具宿存花柱。花期 7~10 月。

Evergreen shrubs or small trees. Palmately compound leaf, leaflets 5–7, thinly leathery, massive, abaxially sparsely stellate pubescent on veins, margin entire, revolute, petiole up to 1 m long. Fruit ovate, with persistent style. Fl. Jul.–Oct..

植株 Whole plant
摄影：黄俞淞 Photo by：Huang Yusong

果序 Infructescence
摄影：黄俞淞 Photo by：Huang Yusong

复叶 Compound leaf
摄影：黄俞淞 Photo by：Huang Yusong

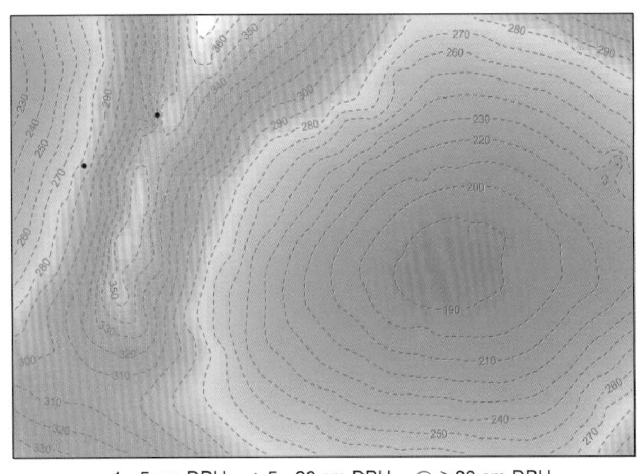

● 1~5 cm DBH　+ 5~20 cm DBH　○ ≥20 cm DBH
个体分布图 Distribution of individuals

径级分布表 DBH class

胸径区间 (Diameter class) (cm)	个体数 (No. of individuals in the plot)	比例 (Proportion) (%)
1~2	1	50.00
2~5	1	50.00
5~10	0	0.00
10~20	0	0.00
20~35	0	0.00
35~60	0	0.00
≥60	0	0.00

168 刺通草 cì tōng cǎo | Himalayan Trevesia

Trevesia palmate (Roxb.) Vis.
五加科 Araliaceae

代码（SpCode）= TREPAL
个体数（Individual number/15 hm^2）= 2
最大胸径（Max DBH）= 3.2 cm
重要值排序（Importance value rank）= 197

常绿小乔木，高 3~8 m。小枝淡黄棕色，具绒毛和刺。叶为单叶，叶片直径达 60~90 cm，革质，掌状深裂，裂片 5~9，披针形，边缘有大锯齿，两面疏生星状绒毛。圆锥花序长约 50 cm，主轴和分枝幼时有锈色绒毛，后毛渐脱落。果实卵球形，直径 1.2~1.8 cm。果期 5~7 月。

Evergreen small trees, 3–8 m tall. Branchlets yellowish brown, with villus and prick. Leaves simple, 60–90 cm in diam. of leaves, leathery, palmate crack, lobes 5–9, lanceolate, margin with sawtooth, both surfaces sparsely stellate pubescent. Panicle ca. 50 cm long, axis and branch rust colored villus when young, glabrescent. Fruit ovate-globose, 1.2–1.8 cm in diam. Fr. May–Jul..

植株 Whole plant
摄影：黄俞淞 Photo by：Huang Yusong

茎 Stem
摄影：黄俞淞 Photo by：Huang Yusong

叶 Leaf
摄影：黄俞淞 Photo by：Huang Yusong

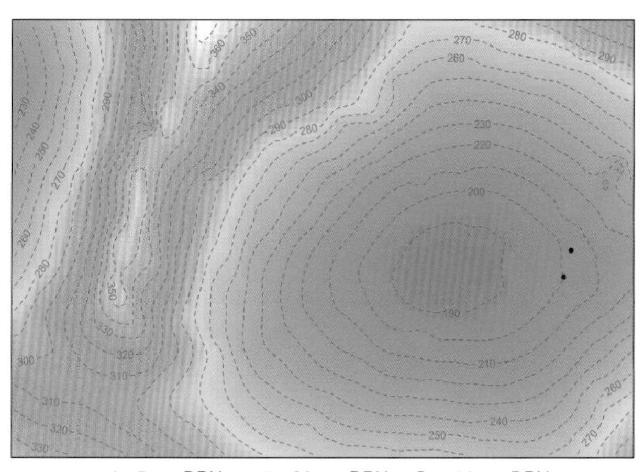

• 1~5 cm DBH + 5~20 cm DBH ○ ≥20 cm DBH
个体分布图 Distribution of individuals

径级分布表 DBH class

胸径区间 (Diameter class) (cm)	个体数 (No. of individuals in the plot)	比例 (Proportion) (%)
1~2	0	0.00
2~5	2	100.00
5~10	0	0.00
10~20	0	0.00
20~35	0	0.00
35~60	0	0.00
≥60	0	0.00

169 湘桂柿

xiāng guì shì | Xianggui Persimmon

Diospyros xiangguiensis S. Lee
柿树科 Ebenaceae

代码（SpCode）= DIOXIA
个体数（Individual number/15 hm^2）= 33
最大胸径（Max DBH）= 26.2 cm
重要值排序（Importance value rank）= 137

灌木或乔木。嫩枝、叶柄、花序和果柄等均密被黄棕色绒毛。叶椭圆形或卵状椭圆形，长 5~10 cm，宽 1.7~4 cm，基部宽楔形。雄花 3~5 朵集成聚伞花序，生于嫩枝下端。雌花单生。果近球形，直径约 1.8 cm，被黄棕色绒毛。花期 6~7 月，果期 9~11 月。

Shrubs or trees. Branchlets, petiole, cymes and fruit petiole densely yellowish tomentose. Leaf blade elliptic or ovate-elliptic, 5–10 cm long, and 1.7–4 cm wide, base broadly cuneate. Male flowers 3–5-flowered cymes, born in branchlets bottom. Female flowers solitary. Berries nearly globose, ca. 1.8 cm in diam., yellowish tomentose. Fl. Jun.–Jul., fr. Sep.–Nov..

果枝 Fruiting branches
摄影：黄俞淞 Photo by：Huang Yusong

叶 Leaves
摄影：黄俞淞 Photo by：Huang Yusong

果 Fruits
摄影：黄俞淞 Photo by：Huang Yusong

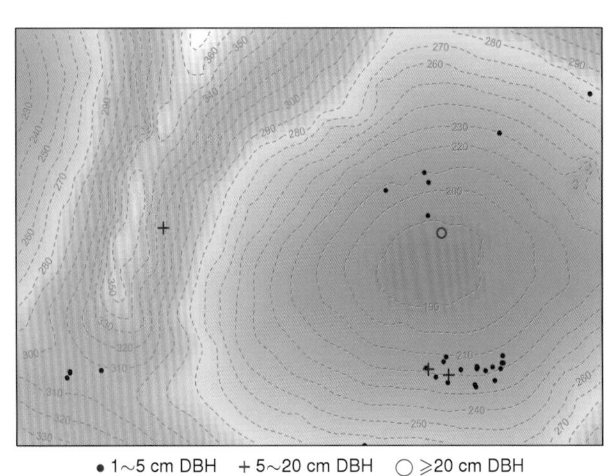

• 1~5 cm DBH　+ 5~20 cm DBH　○ ≥20 cm DBH
个体分布图 Distribution of individuals

径级分布表 DBH class

胸径区间 (Diameter class) (cm)	个体数 (No. of individuals in the plot)	比例 (Proportion) (%)
1~2	11	33.33
2~5	18	54.55
5~10	3	9.09
10~20	0	0.00
20~35	1	3.03
35~60	0	0.00
≥60	0	0.00

170 乌材

wū cái | Erianthous Persimmon

Diospyros eriantha Champ. ex Benth.
柿树科 Ebenaceae

代码（SpCode）= DIOERI
个体数（Individual number/15 hm^2）= 796
最大胸径（Max DBH）= 30.2 cm
重要值排序（Importance value rank）= 24

常绿乔木或灌木，高可达 16 m。树皮灰色，灰褐色至黑褐色。叶纸质，长圆状披针形，边缘具缘毛，微背卷。聚伞花序，腋生。果卵形或长圆形，宿存萼增大，4 裂。花期 7~8 月，果期 10 月至翌年 2 月。

Evergreen trees or shrubs, up to 16 m tall. Bark gray, grayish brown to blackish brown. Leaves papery, oblong-lanceolate, margin with strigose, slightly reflexed. Cyme axillary. Fruit ovate or oblong, persistent calyx spreading, 4-lobed. Fl. Jul.–Aug., fr. Oct.–Feb. of next year.

树干 Trunk
摄影：王斌 Photo by：Wang Bin

果枝 Fruiting branch
摄影：黄俞淞 Photo by：Huang Yusong

叶背 Leaf back
摄影：丁涛 Photo by：Ding Tao

• 1~5 cm DBH　+ 5~20 cm DBH　○ ⩾20 cm DBH

个体分布图 Distribution of individuals

径级分布表 DBH class

胸径区间 (Diameter class) (cm)	个体数 (No. of individuals in the plot)	比例 (Proportion) (%)
1~2	290	36.43
2~5	333	41.83
5~10	143	17.96
10~20	27	3.39
20~35	3	0.38
35~60	0	0.00
⩾60	0	0.00

171 石山柿

shí shān shì | Rockdwelling Persimmon

Diospyros saxatilis S. Lee
柿树科 Ebenaceae

代码（SpCode）= DIOSAX
个体数（Individual number/15 hm^2）= 99
最大胸径（Max DBH）= 10.6 cm
重要值排序（Importance value rank）= 100

落叶小乔木，高 1~5 m。叶薄革质，披针形、卵状披针形或倒披针形，有短尖头，边缘微背卷。果卵形，直径约 1 cm，成熟时由黄色变红色以至紫黑色，顶端具小尖头，宿存萼裂片狭长三角形。花期 4~5 月，果期 10 月至翌年 2 月。

Deciduous small trees, 1–5 m tall. Leaves thinly leathery, lanceolate, ovate-lanceolate or oblanceolate, with mucronate, margin slightly reflexed. Fruit ovate, ca. 1 cm in diam., yellow to red when mature, finally purple black, apex with mucronate, persistent calyx narrow triangle. Fl. Apr.–May, fr. Oct.–Feb. of next year.

果枝 Fruiting branches
摄影：黄俞淞 Photo by: Huang Yusong

枝叶 Branch and leaves
摄影：黄俞淞 Photo by: Huang Yusong

果 Fruit
摄影：黄俞淞 Photo by: Huang Yusong

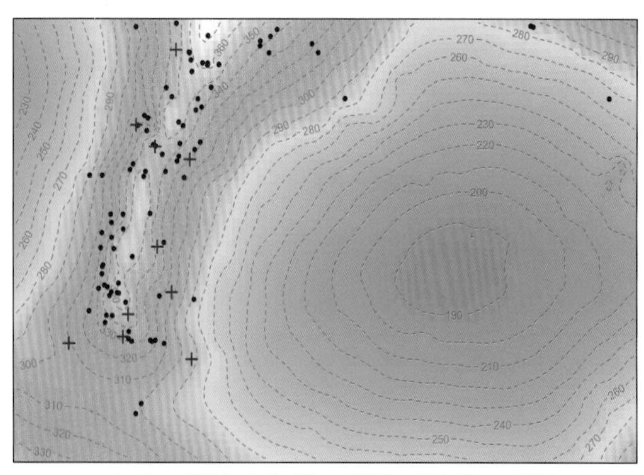
• 1~5 cm DBH + 5~20 cm DBH ○ ≥20 cm DBH
个体分布图 Distribution of individuals

径级分布表 DBH class

胸径区间 (Diameter class) (cm)	个体数 (No. of individuals in the plot)	比例 (Proportion) (%)
1~2	39	39.39
2~5	50	50.51
5~10	8	8.08
10~20	2	2.02
20~35	0	0.00
35~60	0	0.00
≥60	0	0.00

172 山榄叶柿

shān lǎn yè shì | Jungleplum-leaf Persimmon

Diospyros siderophylla H. L. Li
柿树科 Ebenaceae

代码（SpCode）= DIOSID
个体数（Individual number/15 hm^2）= 1592
最大胸径（Max DBH）= 24.9 cm
重要值排序（Importance value rank）= 16

常绿乔木，高可达 15 m。树皮黑褐色。叶近革质，长圆形，叶干时带黑色。雄花腋生，单生或 2 至数朵簇生，无梗。果单生，近无梗，球形，密被棕色短硬伏毛，宿存萼 4 裂。花期 6 月，果期 10~11 月。

Evergreen trees, up to 15 m tall. Bark dark brown. Leaves nearly leathery, oblong, dark when dry. Male flowers axillary, solitary or 2 to several in clusters, sessile. Fruit solitary, subsessile, globose, densely shortly appressed hirsute, persistent calyx 4-lobes. Fl. Jun., fr. Oct.–Nov..

树干 Trunk
摄影：王斌 Photo by：Wang Bin

果 Fruits
摄影：黄俞淞 Photo by：Huang Yusong

叶 Leaves
摄影：王斌 Photo by：Wang Bin

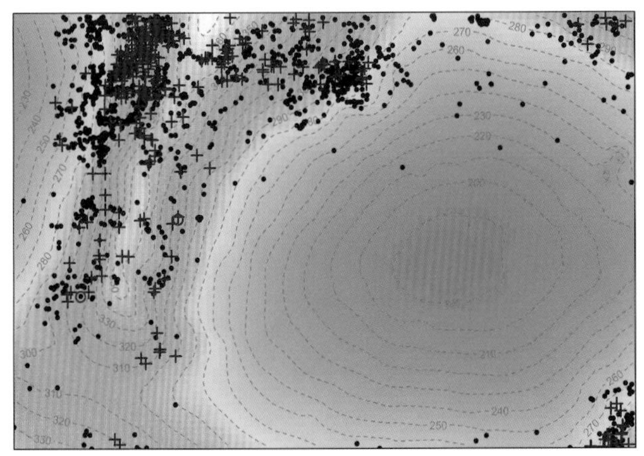

• 1~5 cm DBH + 5~20 cm DBH ○ ≥20 cm DBH
个体分布图 Distribution of individuals

径级分布表 DBH class

胸径区间 (Diameter class) (cm)	个体数 (No. of individuals in the plot)	比例 (Proportion) (%)
1~2	784	49.25
2~5	516	32.41
5~10	166	10.43
10~20	123	7.73
20~35	3	0.19
35~60	0	0.00
≥60	0	0.00

173 金叶树

jīn yè shù | Manyflower Starapple

Chrysophyllum lanceolatum var. *stellatocarpon* P. Royen

山榄科 Sapotaceae

代码（SpCode）= CHRLAN

个体数（Individual number/15 hm^2）= 1

最大胸径（Max DBH）= 6.8 cm

重要值排序（Importance value rank）= 202

乔木或小乔木，高 10~20 m。小枝圆柱形，被黄色柔毛。叶坚纸质，长圆状披针形，先端通常渐尖或尾尖，基部稍偏斜，边缘波状，叶柄被锈色短柔毛。花数朵簇生叶腋。果近球形，具 5 棱，幼时被锈色柔毛。花期 5 月，果期 10 月。

Trees or small trees, 10–20 m tall. Branchlets terete, yellow pubescent. Leaves thickly papery,oblong-lanceolate, apex usually acuminate or caudate, base slightly oblique, margin wavy, petioles rust colored pubescent. Flowers several, fascicled, axillary. Fruit subglobose, 5-ribbed, rust colored tomentose when young. Fl. May, fr. Oct..

果枝 Fruiting branch

摄影：黄俞淞 Photo by：Huang Yusong

果 Fruits

摄影：黄俞淞 Photo by：Huang Yusong

叶 Leaf

摄影：黄俞淞 Photo by：Huang Yusong

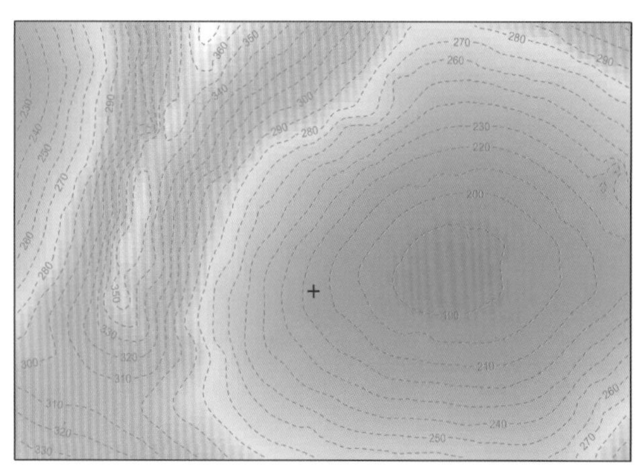

• 1~5 cm DBH + 5~20 cm DBH ○ ≥20 cm DBH

个体分布图 Distribution of individuals

径级分布表 DBH class

胸径区间 (Diameter class) (cm)	个体数 (No. of individuals in the plot)	比例 (Proportion) (%)
1~2	0	0.00
2~5	0	0.00
5~10	1	100.00
10~20	0	0.00
20~35	0	0.00
35~60	0	0.00
≥60	0	0.00

174 紫荆木

zǐ jīng mù | Pasquier's Madhuca

Madhuca pasquieri (Dubard) H. J. Lam
山榄科 Sapotaceae

代码（SpCode）＝ MADPAS
个体数（Individual number/15 hm^2）＝ 34
最大胸径（Max DBH）＝ 21.2 cm
重要值排序（Importance value rank）＝ 118

常绿乔木，高可达 30 m。树皮灰黑色，具乳汁。叶互生，星散或密聚于分枝顶端，革质，倒卵形或倒卵状长圆形，侧脉成 13~26 对。花数朵簇生叶腋，被锈色或灰色短柔毛。果椭圆形或小球形，先端具宿存、花后延长的花柱。花期 7~9 月，果期 10 月至翌年 1 月。

Evergreen trees, up to 30 m tall. Bark blackish, with latex. Leaves alternate, scattered or more often closely clustered at end of branchlets, leathery, obovate or obovate-oblong, lateral veins 13–26 pairs. Flowers several, axillary, fascicled, rust colored to grayish pubescent. Fruit elliptic or globose, apex with lengthened style. Fl. Jul.–Sep., fr. Oct.–Jan. of next year.

果枝 Fruiting branch
摄影：黄俞淞 Photo by：Huang Yusong

果 Fruit
摄影：黄俞淞 Photo by：Huang Yusong

嫩叶 New Leaves
摄影：丁涛 Photo by：Ding Tao

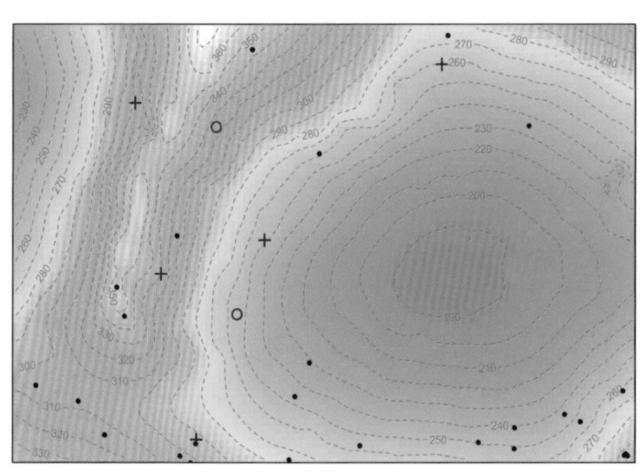

● 1~5 cm DBH ＋ 5~20 cm DBH ○ ⩾20 cm DBH
个体分布图 Distribution of individuals

径级分布表 DBH class

胸径区间 (Diameter class) (cm)	个体数 (No. of individuals in the plot)	比例 (Proportion) (%)
1~2	13	38.24
2~5	14	41.18
5~10	4	11.76
10~20	1	2.94
20~35	2	5.88
35~60	0	0.00
⩾60	0	0.00

175 铁榄

tiě lǎn | Peduncled Sinosideroxylon

Sinosideroxylon pedunculatum (Hemsl.) H. Chuang
山榄科 Sapotaceae

代码（SpCode）= SINPED
个体数（Individual number/15 hm^2）= 8
最大胸径（Max DBH）= 20.4 cm
重要值排序（Importance value rank）= 166

乔木或小乔木，高 (5)9~12 m。小枝圆柱形，被锈色柔毛, 具皮孔。叶互生，密聚小枝先端，革质，卵形或卵状披针形，两面无毛，叶柄被锈色绒毛或近无毛。花浅黄色，1~3 朵簇生于腋生的花序梗上，组成总状花序。浆果卵球形，具花后延长的花柱。花期 5~8 月。

Trees or small trees, (5)9–12 m tall. Branchlets terete, rust colored pubescent and lenticellate. Leaves alternate, often closely clustered at end of branchlets, ovate or ovate-lanceolate, both surfaces glabrous, petiole rust colored pubescent or nearly glabrous. Flowers yellowish, 1–3-flowered clusters on axillary raceme. Berry ovate-globose, with lengthened style. Fl. May–Aug..

植株 Whole plant
摄影：黄俞淞 Photo by: Huang Yusong

花序 Inflorescence
摄影：黄俞淞 Photo by: Huang Yusong

叶 Leaf
摄影：黄俞淞 Photo by: Huang Yusong

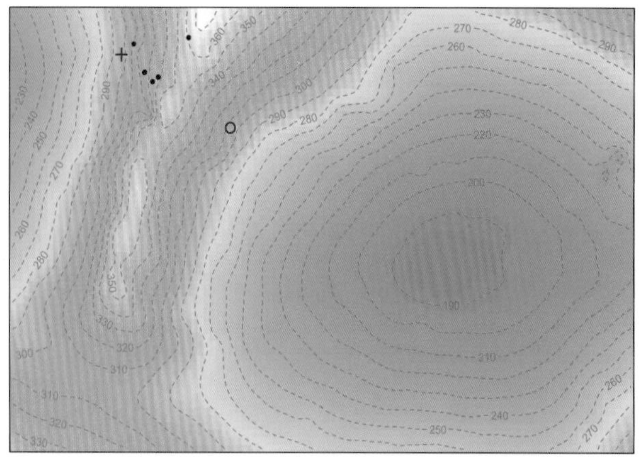

• 1~5 cm DBH + 5~20 cm DBH ○ ≥20 cm DBH
个体分布图 Distribution of individuals

径级分布表 DBH class

胸径区间 (Diameter class) (cm)	个体数 (No. of individuals in the plot)	比例 (Proportion) (%)
1~2	5	62.50
2~5	1	12.50
5~10	0	0.00
10~20	1	12.50
20~35	1	12.50
35~60	0	0.00
≥60	0	0.00

176 毛叶铁榄

máo yè tiě lǎn | Hairy-leaf Peduncled Sinosideroxylon

Sinosideroxylon pedunculatum var. *pubifolium* H. Chuang
山榄科 Sapotaceae

代码（SpCode）= SINPUB
个体数（Individual number/15 hm^2）= 366
最大胸径（Max DBH）= 24.0 cm
重要值排序（Importance value rank）= 38

常绿灌木或乔木，高 5~12 m。小枝圆柱形，被锈色柔毛。叶互生，密聚小枝先端，革质，卵形或卵状披针形，嫩叶两面及小枝先端密被锈色绢毛，后叶面变无毛，叶背绢毛变褐色。花 1~3 朵簇生于腋生的花序梗上，总状花序。浆果卵球形。花期 7~8 月。

Evergreen shrubs or trees, 5–12 m tall. Branchlets terete, rust colored pubescent. Leaves alternate, clusters at end of branchlet, leathery, ovate or ovate-lanceolate, both surfaces of lender leaf and apex of branchlet densely rust colored silk, adaxially glabrous when old, abaxially silk becoming brown. 1–3-flowered clusters on axillary raceme. Berry ovate-globose. Fl. Jul.–Aug..

树干 Trunk
摄影：王斌 Photo by：Wang Bin

果 Fruits
摄影：黄俞淞 Photo by：Huang Yusong

枝叶 Branch and leaves
摄影：黄俞淞 Photo by：Huang Yusong

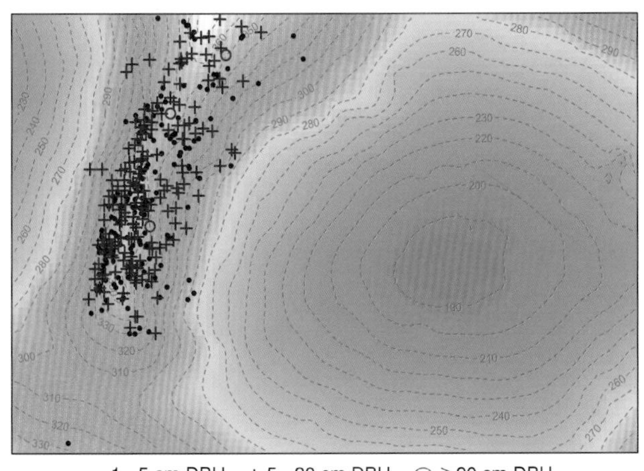

• 1~5 cm DBH + 5~20 cm DBH ○ ≥20 cm DBH
个体分布图 Distribution of individuals

径级分布表 DBH class

胸径区间 (Diameter class) (cm)	个体数 (No. of individuals in the plot)	比例 (Proportion) (%)
1~2	51	13.93
2~5	126	34.43
5~10	119	32.51
10~20	67	18.31
20~35	3	0.82
35~60	0	0.00
≥60	0	0.00

177 块根紫金牛

kuài gēn zǐ jīn niú | Tuberous Ardisia

Ardisia pseudocrispa Pit.
紫金牛科 Myrsinaceae

代码（SpCode）= ARDPSE
个体数（Individual number/15 hm^2）= 2
最大胸径（Max DBH）= 1.7 cm
重要值排序（Importance value rank）= 194

灌木, 高 1~3 m。植株下部具块根，小枝无毛或有时被微柔毛。叶片椭圆形或倒卵状披针形, 长 5~8 cm, 宽 1.5~2.5 cm, 两面无毛。复伞形花序, 花梗长约 1 cm。果球形, 直径约 8 毫 mm, 鲜红色, 具腺点。花期 4~5 月, 果期 11~12 月。

Shrubs, 1–3 m tall. With earthnut, branchlets glabrous or sometimes pubescent. Leaf blade elliptic or obovate-lanceolate, 5–8 cm long, 1.5–2.5 cm wide, both surfaces glabrous. Compounded umbel, pedicel ca. 1 cm long. Fruit globose, ca. 8 mm in diam., scarlet, with gland spot. Fl. Apr.–May, fr. Nov.–Dec..

植株 Whole plant
摄影：黄俞淞 Photo by：Huang Yusong

花序 Inflorescence
摄影：黄俞淞 Photo by：Huang Yusong

叶 Leaves
摄影：黄俞淞 Photo by：Huang Yusong

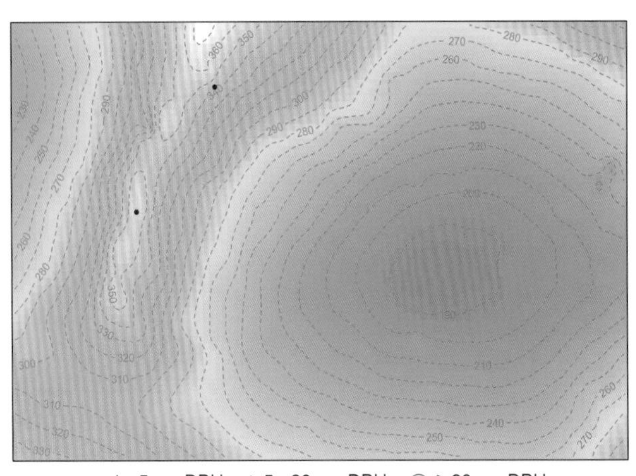

• 1~5 cm DBH + 5~20 cm DBH ○ ≥20 cm DBH
个体分布图 Distribution of individuals

径级分布表 DBH class

胸径区间 (Diameter class) (cm)	个体数 (No. of individuals in the plot)	比例 (Proportion) (%)
1~2	2	100.00
2~5	0	0.00
5~10	0	0.00
10~20	0	0.00
20~35	0	0.00
35~60	0	0.00
≥60	0	0.00

178 南方紫金牛 nán fāng zǐ jīn niú | Thyrsiflorous Ardisia

Ardisia thyrsiflora D. Don
紫金牛科 Myrsinaceae

代码（SpCode）= ARDTHY
个体数（Individual number/15 hm^2）= 448
最大胸径（Max DBH）= 5.0 cm
重要值排序（Importance value rank）= 47

常绿灌木或小乔木，高 1.5~5 m。嫩枝、叶柄、花序均密被锈色微柔毛。叶全缘，侧脉多于 20 对，几与中脉成直角，不连成边缘脉。复亚伞形花序组成圆锥花序，侧生或顶生。果球形，紫红色。花期 3~5 月，果期 10~12 月。

Evergreen shrubs or small trees, 1.5–5 m tall. Branchlets, petioles and inflorescences densely rust pubescent. Leaf blade entire, lateral veins more than 20 pairs, nearly at right angle wiht the midvein, marginal vein absent. Inflorescences terminal or subterminal, paniculate, branches corymbose. Fruit globose, purplish red. Fl. Mar.–May, fr. Oct.–Dec..

果枝 Fruiting branch
摄影：黄俞淞 Photo by：Huang Yusong

叶 Leaves
摄影：王斌 Photo by：Wang Bin

果 Fruits
摄影：黄俞淞 Photo by：Huang Yusong

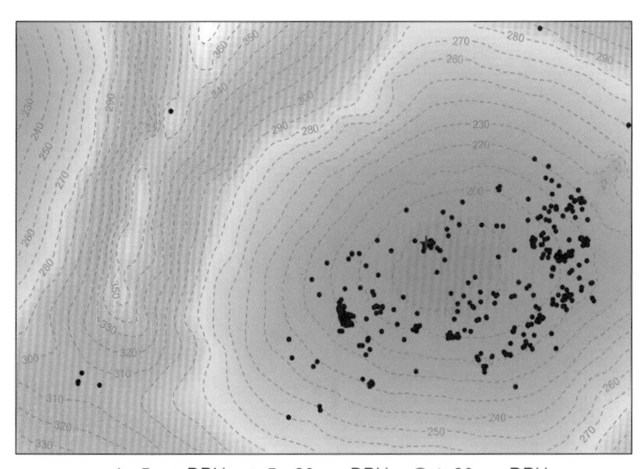

• 1~5 cm DBH + 5~20 cm DBH ○ ≥20 cm DBH

个体分布图 Distribution of individuals

径级分布表 DBH class

胸径区间 (Diameter class) (cm)	个体数 (No. of individuals in the plot)	比例 (Proportion) (%)
1~2	290	64.73
2~5	157	35.04
5~10	1	0.22
10~20	0	0.00
20~35	0	0.00
35~60	0	0.00
≥60	0	0.00

179 中越杜茎山

zhōng yuè dù jīng shān | Balansa's Maesa

Maesa balansae Mez
紫金牛科 Myrsinaceae

代码（SpCode）= MAEBAL
个体数（Individual number/15 hm^2）= 322
最大胸径（Max DBH）= 11.6 cm
重要值排序（Importance value rank）= 42

常绿灌木，高 1~3 m。小枝圆柱形，红褐色。叶片坚纸质，广椭圆形或椭圆状卵形，近全缘或具疏细齿或短锐齿，齿尖常具腺点。圆锥花序，腋生和顶生，花冠白色；果球形。花期 1~2 月，果期 8~11 月。

Evergreen shrubs, 1–3 m tall. Branchlets terete, red brown. Leaf blade stiffly papery, broadly elliptic or elliptic-ovate, nearly entire or serrate, prong usually with dots. Panicle axillary or terminate, corolla white, fruit globose. Fl. Jan.–Feb., fr. Aug.–Nov..

花枝 Flowering branch
摄影：黄俞淞 Photo by：Huang Yusong

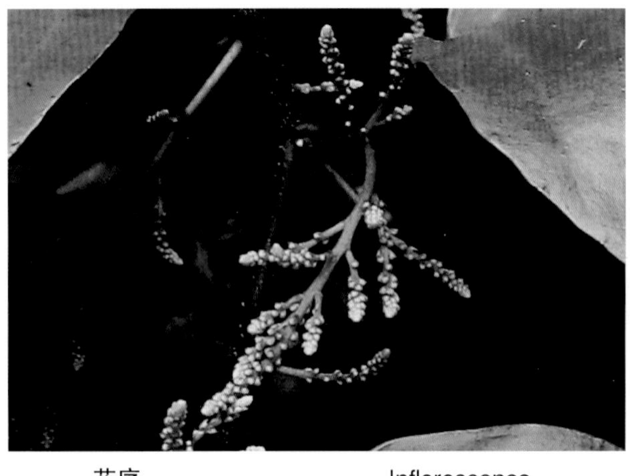

花序 Inflorescence
摄影：黄俞淞 Photo by：Huang Yusong

果序 Infructescence
摄影：黄俞淞 Photo by：Huang Yusong

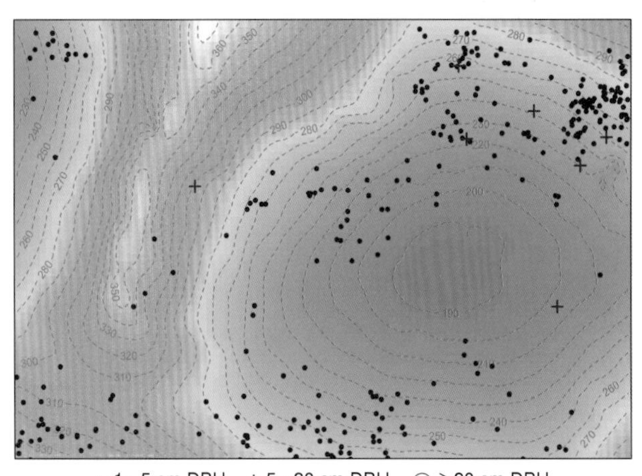

• 1~5 cm DBH + 5~20 cm DBH ○ ≥20 cm DBH
个体分布图 Distribution of individuals

径级分布表 DBH class

胸径区间 (Diameter class) (cm)	个体数 (No. of individuals in the plot)	比例 (Proportion) (%)
1~2	147	45.65
2~5	168	52.17
5~10	6	1.86
10~20	1	0.31
20~35	0	0.00
35~60	0	0.00
≥60	0	0.00

180 鲫鱼胆

jī yú dǎn | Pearly-shining Maesa

Maesa perlarius (Lour.) Merr.
紫金牛科 Myrsinaceae

代码（SpCode）= MAEPER
个体数（Individual number/15 hm^2）= 182
最大胸径（Max DBH）= 8.8 cm
重要值排序（Importance value rank）= 66

常绿小灌木，高 1~3 m。小枝被长硬毛或短柔毛。叶广椭圆状卵形至椭圆形，边缘从中下部以上具粗锯齿，下部全缘，侧脉尾端直达齿尖。总状花序或圆锥花序，腋生，被长硬毛和短柔毛。果球形，无毛。花期 3~4 月，果期 12 月至翌年 5 月。

Evergreen small shrubs, 1–3 m tall. Branchlets hirtellous or pubescence. Leaf blade broadly ovtae to elliptic, margin coarsely serrate distally and entire toward base, lateral veins ending in teeth. Raceme or panicle, axillary, hirtellous and pubescence. Fruit globose, glabrous. Fl. Mar.–Apr., fr. Dec.–May of next year.

叶 Leaves
摄影：黄俞淞 Photo by: Huang Yusong

花序 Inflorescence
摄影：黄俞淞 Photo by: Huang Yusong

果序 Infructescence
摄影：黄俞淞 Photo by: Huang Yusong

• 1~5 cm DBH + 5~20 cm DBH ○ ≥20 cm DBH
个体分布图 Distribution of individuals

径级分布表 DBH class

胸径区间 (Diameter class) (cm)	个体数 (No. of individuals in the plot)	比例 (Proportion) (%)
1~2	91	50.00
2~5	82	45.05
5~10	9	4.95
10~20	0	0.00
20~35	0	0.00
35~60	0	0.00
≥60	0	0.00

181 密花树

mì huā shù | Nerium-leaf Rapanea

Myrsine seguinii H. Lév.

紫金牛科 Myrsinaceae

代码（SpCode）= MYRSEG

个体数（Individual number/15 hm^2）= 2

最大胸径（Max DBH）= 2.2 cm

重要值排序（Importance value rank）= 199

灌木或小乔木，高 2~7 m。小枝无毛。叶片革质，长圆状倒披针形至倒披针形，基部楔形，多少下延，全缘，两面无毛。伞形花序，花簇生，有花 3~10 朵。果球形或近卵形，灰绿色或紫黑色。花期 4~5 月，果期 10~12 月。

Shrubs or small trees, 2–7 m tall. Branchlets glabrous. Leaves leathery, oblong-oblanceolate to oblanceolate, base cuneate, decurrent, margin entire, both surfaces glabrous. Umbel, 3–10-flowered, clustered. Fruits globose or subovate, grayish green or purple black. Fl. Apr.–May, fr. Oct.–Dec..

花枝 Flowering branches

摄影：黄俞淞 Photo by：Huang Yusong

叶 Leaves

摄影：黄俞淞 Photo by：Huang Yusong

果 Fruits

摄影：黄俞淞 Photo by：Huang Yusong

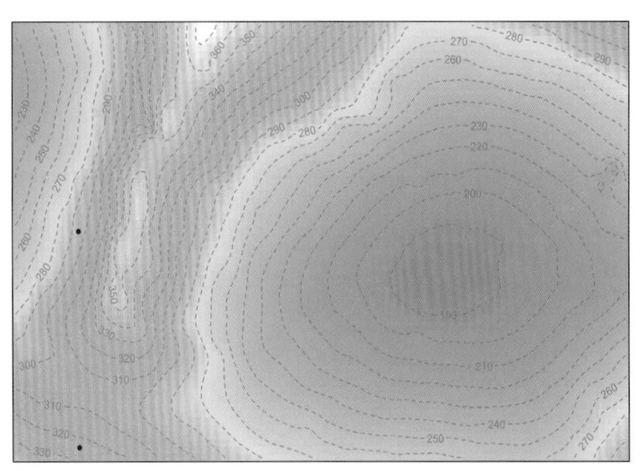

• 1~5 cm DBH　+ 5~20 cm DBH　○ ≥20 cm DBH

个体分布图 Distribution of individuals

径级分布表 DBH class

胸径区间 (Diameter class) (cm)	个体数 (No. of individuals in the plot)	比例 (Proportion) (%)
1~2	0	0.00
2~5	2	100.00
5~10	0	0.00
10~20	0	0.00
20~35	0	0.00
35~60	0	0.00
≥60	0	0.00

182 广西流苏树

guǎng xī liú sū shù | Guangxi Fringe-tree

Chionanthus guangxiensis B.M. Miao
木犀科 Oleaceae

代码（SpCode）= CHIGUA
个体数（Individual number/15 hm^2）= 198
最大胸径（Max DBH）= 13.3 cm
重要值排序（Importance value rank）= 98

常绿灌木或小乔木，高 3~6 m。小枝暗灰色。叶革质，椭圆形或窄椭圆形，叶基部变狭成一有翅的叶柄，全缘，两面光滑无毛。圆锥花序腋生，被微柔毛。果椭圆形，被白粉。花期 4 月，果期 8 月。

Evergreen shrubs or small trees, 3–6 m tall. Branchlets dark gray. Leaves leathery, elliptic or narrowly elliptic, base attenuate and decurrent into a winged petiole, margin entire, both surfaces glabrous. Panicle axillary, pubescent. Fruits elliptic, pruinose. Fl. Apr., fr. Aug..

枝叶 Branch and leaves
摄影：黄俞淞 Photo by：Huang Yusong

花序 Inflorescence
摄影：黄俞淞 Photo by：Huang Yusong

果枝 Fruiting branch
摄影：黄俞淞 Photo by：Huang Yusong

• 1~5 cm DBH + 5~20 cm DBH ○ ≥20 cm DBH
个体分布图 Distribution of individuals

径级分布表 DBH class

胸径区间 (Diameter class) (cm)	个体数 (No. of individuals in the plot)	比例 (Proportion) (%)
1~2	66	33.33
2~5	111	56.06
5~10	20	10.10
10~20	1	0.51
20~35	0	0.00
35~60	0	0.00
≥60	0	0.00

183 枝花流苏树

zhī huā liú sū shù | Ramiflorous Fringe-tree

Chionanthus ramiflorus Roxb.

木犀科 Oleaceae

代码（SpCode）＝ CHIRAM

个体数（Individual number/15 hm^2）＝ 354

最大胸径（Max DBH）＝ 42.1 cm

重要值排序（Importance value rank）＝ 44

常绿灌木或乔木，高 3~25 m。树皮灰黑色或灰褐色，小枝灰白色或褐色，紫红色，节间常压扁。叶片椭圆形、长圆状椭圆形或卵状椭圆形。花序腋生，稀顶生。果卵状椭圆形或椭圆形，呈蓝黑色，被白粉。花期 12 月至翌年 6 月，果期 5 月至翌年 3 月。

Evergreen shrubs or trees, up to 3–25 m tall. Bark gray black or gray-brown, branchlets white or brown, pureple, usually compressed. Leaf blade elliptic, oblong-elliptic or ovate-elliptic. Panicles axillary or rarely terminal; Drupeovate-elliptic or elliptic, blue-black, pruinose. Fl. Dec.–Jun. of next year, fr. May–Mar. of next year.

果枝 Fruiting branches

摄影：黄俞淞 Photo by: Huang Yusong

果序 Infructescence

摄影：黄俞淞 Photo by: Huang Yusong

枝叶 Branch and leaves

摄影：黄俞淞 Photo by: Huang Yusong

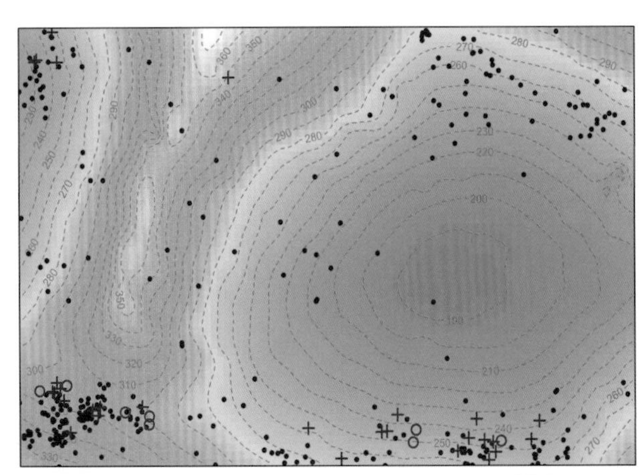

• 1~5 cm DBH　+ 5~20 cm DBH　○ ≥20 cm DBH

个体分布图 Distribution of individuals

径级分布表 DBH class

胸径区间 (Diameter class) (cm)	个体数 (No. of individuals in the plot)	比例 (Proportion) (%)
1~2	213	60.17
2~5	104	29.38
5~10	14	3.95
10~20	15	4.24
20~35	7	1.98
35~60	1	0.28
≥60	0	0.00

184 糖胶树

táng jiāo shù | Common Alstonia

Alstonia scholaris（*L.*）*R.Br.*
夹竹桃科 Apocynaceae

代码（SpCode）= ALSSCH
个体数（Individual number/15 hm^2）= 2
最大胸径（Max DBH）= 36.0 cm
重要值排序（Importance value rank）= 172

常绿乔木，高达 40 m。树皮灰色，树干具乳汁。叶 3~10 片轮生，倒卵状长圆形、倒披针形或匙形。侧脉密生而平行。花白色，多朵组成稠密的聚伞花序，顶生，被柔毛。蓇葖 2，细长，线性。花期 6~11 月，果期 10~12 月。

Evergreen trees, up to 40 m tall. Bark gray, trunk with milky juice. Leaves in whorls of 3–10, obovate-oblong, oblanceolate or cochlear. Lateral veins densely and parallel. cymes dense, with many white flowers, terminal, pubescent. Follicles 2, linear. Fl. Jun.–Nov., fr. Oct.–Dec..

植株 Whole plant
摄影：黄俞淞 Photo by：Huang Yusong

果 Fruits
摄影：黄俞淞 Photo by：Huang Yusong

复叶 Compound leaves
摄影：黄俞淞 Photo by：Huang Yusong

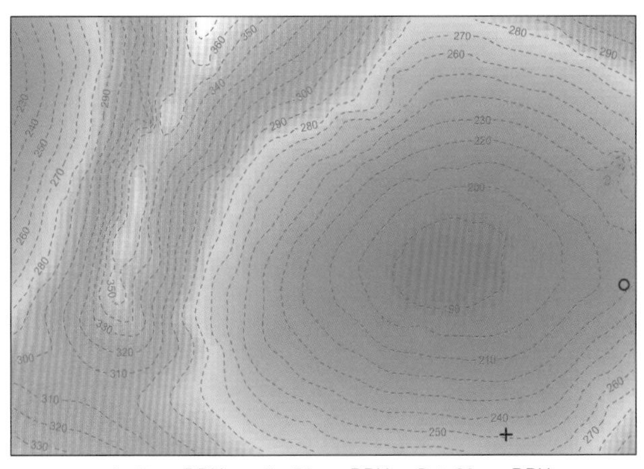
• 1~5 cm DBH + 5~20 cm DBH ○ ≥20 cm DBH
个体分布图 Distribution of individuals

径级分布表 DBH class

胸径区间 (Diameter class) (cm)	个体数 (No. of individuals in the plot)	比例 (Proportion) (%)
1~2	0	0.00
2~5	0	0.00
5~10	0	0.00
10~20	1	50.00
20~35	0	0.00
35~60	1	50.00
≥60	0	0.00

185 云南倒吊笔

yún nán dào diào bǐ | Scarlet Wrightia

Wrightia coccinea (Roxb.) Sims
夹竹桃科 Apocynaceae

代码（SpCode）= WRICOC
个体数（Individual number/15 hm^2）= 129
最大胸径（Max DBH）= 44.4 cm
重要值排序（Importance value rank）= 58

常绿小乔木，高约 8 m。小枝灰色或褐色。叶膜质，椭圆形至卵圆形，无毛或叶背脉上有微毛。花通常单生，或数朵组成顶生聚伞花序，副花冠不分裂，杯状；蓇葖 2 个粘生，无毛，具白色斑点。花期 1~5 月，果期 6~12 月。

Evergreen small trees, ca. 8 m tall. Branchlets gray or brown. Leaves membranous, elliptic, to oval, glabrous or veins sparsely tomentose abaxially. Flowers usually solitary, or cymes terminal. Corona cup-shaped. Follicles 2, connate, glabrous, with white spotted. Fl. Jan.–May, fr. Jun.–Dec..

树干 Trunk
摄影：王斌 Photo by：Wang Bin

花 Flower
摄影：丁涛 Photo by：Ding Tao

叶 Leaves
摄影：黄俞淞 Photo by：Huang Yusong

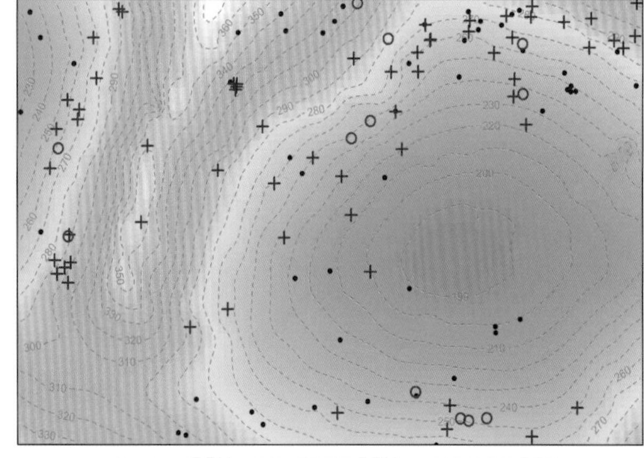

• 1~5 cm DBH ＋5~20 cm DBH ○ ≥20 cm DBH
个体分布图 Distribution of individuals

径级分布表 DBH class

胸径区间 (Diameter class) (cm)	个体数 (No. of individuals in the plot)	比例 (Proportion) (%)
1~2	26	20.16
2~5	29	22.48
5~10	35	27.13
10~20	27	20.93
20~35	11	8.53
35~60	1	0.78
≥60	0	0.00

186 水团花

shuǐ tuán huā | Pilular Adina

Adina pilulifera (Lam.) Franch. ex Drake
茜草科 Rubiaceae

代码（SpCode）＝ ADIPIL
个体数（Individual number/15 hm^2）＝ 4
最大胸径（Max DBH）＝ 13.5 cm
重要值排序（Importance value rank）＝ 179

常绿灌木至小乔木，高达 5 m。叶对生，厚纸质，椭圆形至椭圆状披针形，或倒卵状长圆形，上面无毛，下面无毛或有时被稀疏短柔毛，脉腋窝陷有稀疏的毛。头状花序明显腋生，极稀顶生。果序直径 8~10 mm，种子长圆形，两端有狭翅。花期 6~7 月，果期 7~12 月。

Evergreen shrubs or small trees, up to 5 m tall. Leaves opposite, thickly papery, elliptic to elliptic-lanceolate, or obovate-oblong, adaxially glabrous, abaxially glabrous or sometimes sparsely pubescent, sparsely pubescent at axils vein. Capitulum obviously axillary, rarely terminal. Infructescence 8–10 mm in diam., seeds oblong, narrowly winged at ends. Fl. Jun.–Jul., fr. Jul.–Dec..

果枝 Fruiting branches
摄影：黄俞淞 Photo by：Huang Yusong

果序 Infructescence
摄影：黄俞淞 Photo by：Huang Yusong

叶 Leaves
摄影：黄俞淞 Photo by：Huang Yusong

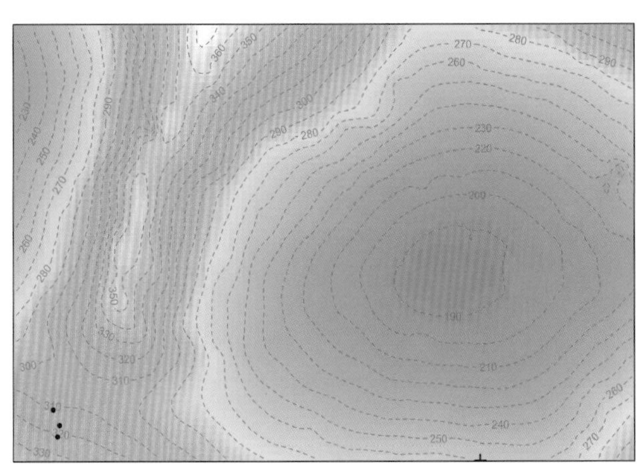
• 1~5 cm DBH　+ 5~20 cm DBH　○ ≥20 cm DBH
个体分布图 Distribution of individuals

径级分布表 DBH class

胸径区间 (Diameter class) (cm)	个体数 (No. of individuals in the plot)	比例 (Proportion) (%)
1~2	1	25.00
2~5	2	50.00
5~10	0	0.00
10~20	1	25.00
20~35	0	0.00
35~60	0	0.00
≥60	0	0.00

187 茜树

qiàn shù | Nambo Aidia

Aidia cochinchinensis Lour.
茜草科 Rubiaceae

代码（SpCode）= AIDCOC
个体数（Individual number/15 hm^2）= 9
最大胸径（Max DBH）= 9.6 cm
重要值排序（Importance value rank）= 161

灌木或乔木，高 2~15 m。枝无毛。叶革质或纸质，对生，长圆状披针形，两面无毛，下面脉腋内的小窝孔中常簇生短柔毛。聚伞花序与叶对生或生于无叶的节上，多花，有短柔毛或无毛。浆果球形，无毛或有疏柔毛，紫黑色。花期 3~6 月，果期 5 月至翌年 2 月。

Shrubs or trees, 2–15 m tall. Branches glabrous. Leaves leathery or papery, opposite, oblong-lanceolate, both surfaces glabrous, in abaxial axils usually with pilosulous or foveolate domatia. Inflorescences cymose, opposite to leaves or borned at leafless joint, many flowers, short pubescent or glabrous. Berry globose, glabrous or sparsely pubescent, purple black. Fl. Mar.–Jun., fr. May–Feb. of next year.

树干 Trunk
摄影：王斌 Photo by: Wang Bin

花枝 Flowering branch
摄影：吴磊 Photo by: Wu Lei

花序 Inflorescence
摄影：吴磊 Photo by: Wu Lei

• 1~5 cm DBH　+ 5~20 cm DBH　○ ≥20 cm DBH

个体分布图 Distribution of individuals

径级分布表 DBH class

胸径区间 (Diameter class) (cm)	个体数 (No. of individuals in the plot)	比例 (Proportion) (%)
1~2	4	44.44
2~5	2	22.22
5~10	3	33.33
10~20	0	0.00
20~35	0	0.00
35~60	0	0.00
≥60	0	0.00

188 鱼骨木

yú gǔ mù | Butulang Canthium

Canthium dicoccum (Gaertn.) Merr.
茜草科 Rubiaceae

代码（SpCode）= CANDIC
个体数（Individual number/15 hm^2）= 897
最大胸径（Max DBH）= 45.1 cm
重要值排序（Importance value rank）= 22

常绿灌木或小乔木，高可达 15 m。小枝嫩时压扁形或四棱柱形，黑褐色。叶革质，叶干后两面极光亮，全缘或微波状。聚伞花序具短的总花梗。核果倒卵形，或倒卵状椭圆形。花期 1~8 月。

Evergreen shrubs or small trees, up to 15 m. Branchlets repressed or quadrangular when young, black brown. Leaves leathery, both very shining when dry, margin entire or slightly wavelike. Cymes with short peduncles. Drupe obovate or obovate-elliptic. Fl. Jan.–Aug..

树干 Trunk
摄影：王斌 Photo by：Wang Bin

花枝 Flowering branches
摄影：黄俞淞 Photo by：Huang Yusong

果序 Infructescence
摄影：刘晟源 Photo by：Liu Shengyuan

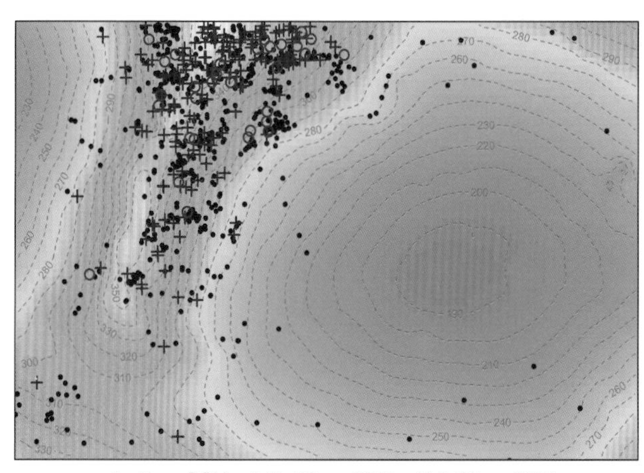

• 1~5 cm DBH + 5~20 cm DBH ○ ≥20 cm DBH
个体分布图 Distribution of individuals

径级分布表 DBH class

胸径区间 (Diameter class) (cm)	个体数 (No. of individuals in the plot)	比例 (Proportion) (%)
1~2	393	43.81
2~5	288	32.11
5~10	100	11.15
10~20	85	9.48
20~35	29	3.23
35~60	2	0.22
≥60	0	0.00

189 大叶鱼骨木

dà yè yú gǔ mù | Large-leaf Canthium

Canthium simile Merr. et Chun
茜草科 Rubiaceae

代码（SpCode）= CANSIM
个体数（Individual number/15 hm^2）= 134
最大胸径（Max DBH）= 29.6 cm
重要值排序（Importance value rank）= 68

常绿直立灌木至小乔木，高 4~10 m。小枝初时微压扁，后呈圆柱形。叶卵状长圆形，托叶基部阔，上部突然收窄成长约 5 mm 的急尖头。花序腋生。核果倒卵形，压扁，孪生。花期 1~3 月，果期 6~7 月。

Evergreen erect shrubs or small tress, 4–10 m tall. Branchlets slightly compressed, becoming terete. Leaves ovate-oblong, stipules basal widely, apex with ca. 5 mm cusp. Inflorescence axillary. Drupe obovate, compressed, twinborn. Fl. Jan.–Mar., fr. Jun.–Jul..

树干 Trunk
摄影：黄俞淞 Photo by：Huang Yusong

花枝 Flowering branch
摄影：黄俞淞 Photo by：Huang Yusong

花序 Inflorescence
摄影：黄俞淞 Photo by：Huang Yusong

• 1~5 cm DBH + 5~20 cm DBH ○ ≥20 cm DBH
个体分布图 Distribution of individuals

径级分布表 DBH class

胸径区间 (Diameter class) (cm)	个体数 (No. of individuals in the plot)	比例 (Proportion) (%)
1~2	36	26.87
2~5	33	24.63
5~10	31	23.13
10~20	27	20.15
20~35	7	5.22
35~60	0	0.00
≥60	0	0.00

190 山石榴

shān shí liú | Mountain Pomegranate

Catunaregam spinosa (Thunb.) Tirveng.
茜草科 Rubiaceae

代码（SpCode） = CATSPI
个体数（Individual number/15 hm^2） = 232
最大胸径（Max DBH） = 15.9 cm
重要值排序（Importance value rank） = 50

常绿灌木或小乔木，高 1~10 m。刺对生，长 1~5 cm。叶对生或簇生于侧生短枝上，倒卵形或长圆状倒卵形，叶背脉腋内常有短束毛。花单生或 2~3 朵簇生于具叶、侧生短枝的顶部；浆果大，球形，直径 2~4 厘米。花期 3~6 月，果期 5 月至翌年 1 月。

Evergreen shrubs or small trees, 1–10 m tall. Paired thorns, 1–5 cm long. Leaves opposite or fascicled on lateral short shoots, obovate ot oblong-obovate, often with pilosulous domatia in abaxial axils. Flowers solitary or 2–3-flowered terminal on lateral short shoots together with tufted leaves. Berry globous, 2–4 cm in diam.. Fl. Mar.–Jun., fr. May–Jan. of next year.

树干 Trunks
摄影：王斌 Photo by：Wang Bin

果枝 Fruiting branch
摄影：黄俞淞 Photo by：Huang Yusong

果 Fruits
摄影：刘晟源 Photo by：Liu Shengyuan

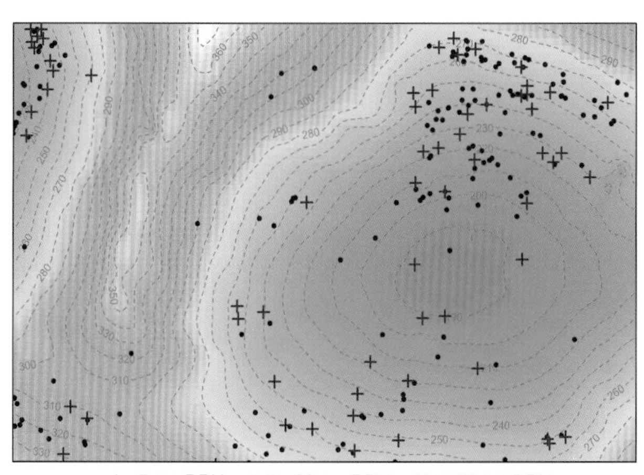

• 1~5 cm DBH + 5~20 cm DBH ○ ≥20 cm DBH
个体分布图 Distribution of individuals

径级分布表 DBH class

胸径区间 (Diameter class) (cm)	个体数 (No. of individuals in the plot)	比例 (Proportion) (%)
1~2	60	25.86
2~5	107	46.12
5~10	60	25.86
10~20	5	2.16
20~35	0	0.00
35~60	0	0.00
≥60	0	0.00

191 土连翘

tǔ lián qiào | Flaccid Hymenodictyon

Hymenodictyon flaccidum Wall.
茜草科 Rubiaceae

代码（SpCode）= HYMFLA
个体数（Individual number/15 hm^2）= 94
最大胸径（Max DBH）= 36.6 cm
重要值排序（Importance value rank）= 86

落叶乔木，高 6~20 m。树皮灰色。叶通常聚生枝顶，顶端骤然短渐尖，叶两面无毛或叶背被疏柔毛，托叶较大，反折。总状花序腋生，被柔毛。蒴果，褐色，有灰白色斑点。花期 5~7 月，果期 8~11 月。

Deciduous trees, 6–20 m tall. Bark gray. Leaves often crowded at ends of branches, apex acute to acuminate or rarely rounded, glabrous on both surfaces or sometimes pilosulous abaxially, stipules reflexed. Inflorescences racemiform, axillary, pubescent. Capsules brown, with several prominent whitened. Fl. May–Jul., fr. Aug.–Nov..

果枝 Fruiting branch
摄影：黄俞淞 Photo by：Huang Yusong

果序 Infructescences
摄影：黄俞淞 Photo by：Huang Yusong

嫩叶 New Leaves
摄影：丁涛 Photo by：Ding Tao

• 1~5 cm DBH + 5~20 cm DBH ○ ≥20 cm DBH
个体分布图 Distribution of individuals

径级分布表 DBH class

胸径区间 (Diameter class) (cm)	个体数 (No. of individuals in the plot)	比例 (Proportion) (%)
1~2	30	31.91
2~5	34	36.17
5~10	16	17.02
10~20	10	10.64
20~35	3	3.19
35~60	1	1.06
≥60	0	0.00

192 白花龙船花 bái huā lóng chuán huā | White-flowered Ixora

Ixora henryi H. Lév.
茜草科 Rubiaceae

代码（SpCode）= IXOHEN
个体数（Individual number/15 hm^2）= 122
最大胸径（Max DBH）= 5.4 cm
重要值排序（Importance value rank）= 96

常绿灌木，高 1~3 m。小枝初时扁圆柱形。叶对生，长圆形或披针形，少近椭圆形，托叶三角形至卵形，近顶部骤然收狭成长 3~11 mm 的芒尖。花序顶生，排成三歧伞房式的聚伞花序。果球形，顶端有细小的萼檐裂片。花期 8~12 月，果期翌年 5~7 月。

Evergreen shrubs, 1–3 m tall. Branches compressed terete. Leaves opposite, oblong ot lanceolate, rarely elliptic, stipules triangular to ovate, shortly obtuse to narrowed and prolonged into arista 3–11 mm. Inflorescences terminal, corymbiform to congested-cymose. Fruit globose, apex with small limb-lobed. Fl. Aug.–Dec., fr. May–Jul. of next year.

花序 Inflorescence
摄影：黄俞淞 Photo by：Huang Yusong

枝叶 Branch and leaves
摄影：黄俞淞 Photo by：Huang Yusong

果枝 Fruiting branch
摄影：黄俞淞 Photo by：Huang Yusong

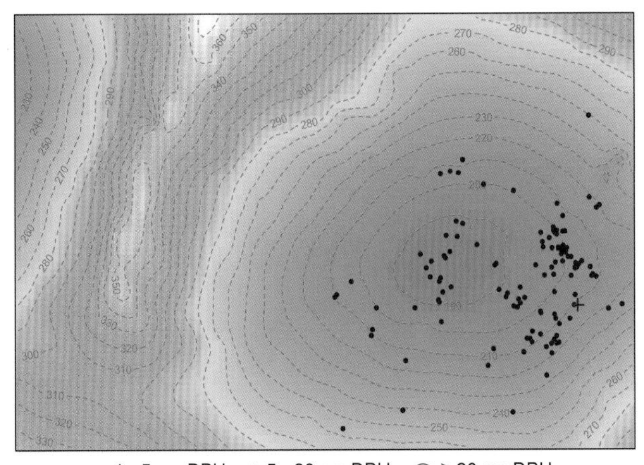

• 1~5 cm DBH + 5~20 cm DBH ○ ≥20 cm DBH
个体分布图 Distribution of individuals

径级分布表 DBH class

胸径区间 (Diameter class) (cm)	个体数 (No. of individuals in the plot)	比例 (Proportion) (%)
1~2	84	68.85
2~5	37	30.33
5~10	1	0.82
10~20	0	0.00
20~35	0	0.00
35~60	0	0.00
≥60	0	0.00

193 鸡仔木

jī zǎi mù | Racemous Sinoadina

Sinoadina racemosa (Sieb. et Zucc.) Ridsdale
茜草科 Rubiaceae

代码（SpCode）= SINRAC
个体数（Individual number/15 hm^2）= 127
最大胸径（Max DBH）= 26.4 cm
重要值排序（Importance value rank）= 81

半常绿或落叶乔木，高 4~12 m。树皮灰色，粗糙。叶对生，薄革质，宽卵形、卵状长圆形或椭圆形。花序密被短柔毛，顶生。小蒴果倒卵状楔形，有稀疏的毛。花、果期 5~12 月。

Semi- to fully deciduous trees, 4–12 m tall. Bark gray, rough. Leaves opposite, thinly leathery, broadly ovate, ovate-oblong ot elliptic. Inflorescence densely puberulent, terminal. Capsules obovate-cuneate, sparsely hirtellous. Fl. and fr. May–Dec..

果枝 Fruiting branches
摄影：黄俞淞 Photo by：Huang Yusong

植株 Whole plant
摄影：黄俞淞 Photo by：Huang Yusong

果 Fruits
摄影：黄俞淞 Photo by：Huang Yusong

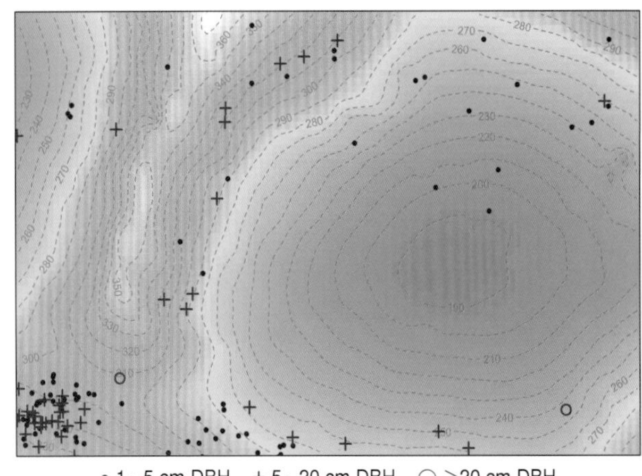

• 1~5 cm DBH + 5~20 cm DBH ○ ≥20 cm DBH
个体分布图 Distribution of individuals

径级分布表 DBH class

胸径区间 (Diameter class) (cm)	个体数 (No. of individuals in the plot)	比例 (Proportion) (%)
1~2	42	33.07
2~5	38	29.92
5~10	29	22.83
10~20	16	12.60
20~35	2	1.57
35~60	0	0.00
≥60	0	0.00

194 假桂乌口树

jiǎ guì wū kǒu shù | Tapered Tarenna

Tarenna attenuata (Voigt) Hutch.
茜草科 Rubiaceae

代码（SpCode）= TARATT
个体数（Individual number/15 hm^2）= 94
最大胸径（Max DBH）= 17.0 cm
重要值排序（Importance value rank）= 92

常绿灌木或小乔木，高 1~8 m。枝灰白色。叶长圆状披针形、长圆状倒卵形、披针形，干时变黑褐色。伞房状的聚伞花序顶生；浆果近球形，成熟时紫黑色，顶部有宿存的花萼，种子 2 颗。花期 4~12 月，果期 5 月至翌年 1 月。

Evergreen shrubs or small trees, 1–8 m tall. Branches gray-white. Leaves oblong-lanceolate, oblong-obovate, lanceolate, black-brown when dry. Corymbose cyme terminal. Berry subglobose, purple black when maturity, calyx persistented on the apex, seeds 2. Fl. Apr.–Dec., fr. May–Jan. of next year.

枝叶 Branch and leaves
摄影：黄俞淞 Photo by：Huang Yusong

花序 Inflorescence
摄影：刘晟源 Photo by：Liu Shengyuan

果枝 Fruiting branch
摄影：黄俞淞 Photo by：Huang Yusong

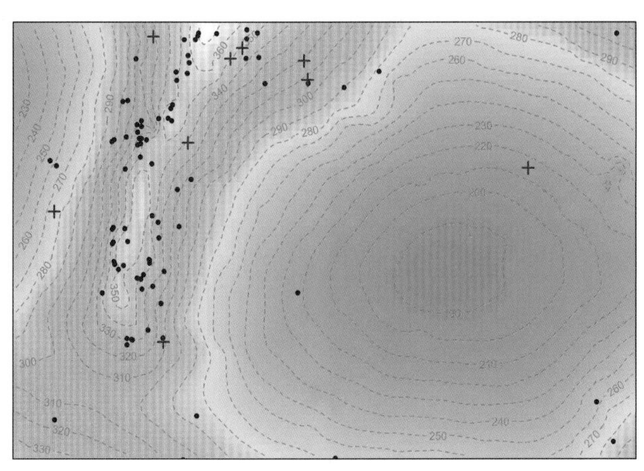

• 1~5 cm DBH + 5~20 cm DBH ○ ≥20 cm DBH
个体分布图 Distribution of individuals

径级分布表 DBH class

胸径区间 (Diameter class) (cm)	个体数 (No. of individuals in the plot)	比例 (Proportion) (%)
1~2	42	44.68
2~5	43	45.74
5~10	6	6.38
10~20	3	3.19
20~35	0	0.00
35~60	0	0.00
≥60	0	0.00

195 龙州水锦树

lóng zhōu shuǐ jǐn shù | Few-Flower Wendlandia

Wendlandia oligantha W. C. Chen
茜草科 Rubiaceae

代码（SpCode）= WENOLI
个体数（Individual number/15 hm^2）= 7
最大胸径（Max DBH）= 7.2 cm
重要值排序（Importance value rank）= 168

灌木或小乔木，高 3~10 m。小枝灰褐色，无毛或有疏短柔毛。叶革质，椭圆形、卵形或卵状长圆形。两面无毛或有时在中脉上有疏短柔毛。托叶顶端圆形，反折。花序顶生，被柔毛。蒴果球形，无毛，直径约 1.5 mm。花期 7~8 月，果期 8~12 月。

Shrubs or small trees, 3–10 m tall. Branchlets grayish brown, glabrous or sparsely pubescent. Leaf blade leathery, elliptic, ovate or ovate-oblong, both surfaces glabrous, or sometimes sparsely pubescent on midvein, stipules reflexed, apex rounded. Inflorescences terminal, pubescent. Capsules globose, glabrous, ca. 1.5 mm in diam.. Fl. Jul.–Aug., fr. Aug.–Dec..

植株 Whole plant
摄影：黄俞淞 Photo by：Huang Yusong

花序 Inflorescences
摄影：黄俞淞 Photo by：Huang Yusong

叶 Leaves
摄影：黄俞淞 Photo by：Huang Yusong

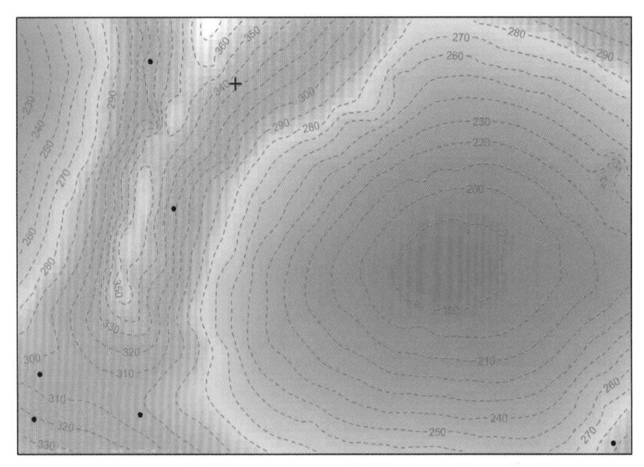

• 1~5 cm DBH + 5~20 cm DBH ○ ≥20 cm DBH
个体分布图 Distribution of individuals

径级分布表 DBH class

胸径区间 (Diameter class) (cm)	个体数 (No. of individuals in the plot)	比例 (Proportion) (%)
1~2	1	14.29
2~5	5	71.43
5~10	1	14.29
10~20	0	0.00
20~35	0	0.00
35~60	0	0.00
≥60	0	0.00

196 水锦树

shuǐ jǐn shù | Uvaria-leaf Wendlandia

Wendlandia uvariifolia Hance
茜草科 Rubiaceae

代码（SpCode）= WENUVA
个体数（Individual number/15 hm^2）= 80
最大胸径（Max DBH）= 21.9 cm
重要值排序（Importance value rank）= 91

灌木或乔木，高 2~15 m。小枝被锈色硬毛。叶纸质，宽椭圆形、卵形或长圆状披针形，上面散生短硬毛，下面密被灰褐色柔毛，叶柄长 0.5~3.5 cm。圆锥状的聚伞花序顶生，被灰褐色硬毛。蒴果小，球形，直径 1~2 mm，被短柔毛。花期 1~5 月，果期 4~10 月。

Shrubs or trees, 2–15 m tall. Branchlets with rust hirsute. Leaf blade papery, oblong, ovate or oblong-lanceolate, adaxially sparsely hirtellous, abaxially densely gray brown pubescent, petiole 0.5–3.5 cm long. Inflorescences paniculate terminal. Densely gray brown hirtellous. Capsules globose, 1–2 mm in diam., pubescent. Fl. Jan.–May, fr. Apr.–Oct..

树干 Trunk
摄影：黄俞淞 Photo by: Huang Yusong

花序 Inflorescences
摄影：黄俞淞 Photo by: Huang Yusong

叶 Leaves
摄影：黄俞淞 Photo by: Huang Yusong

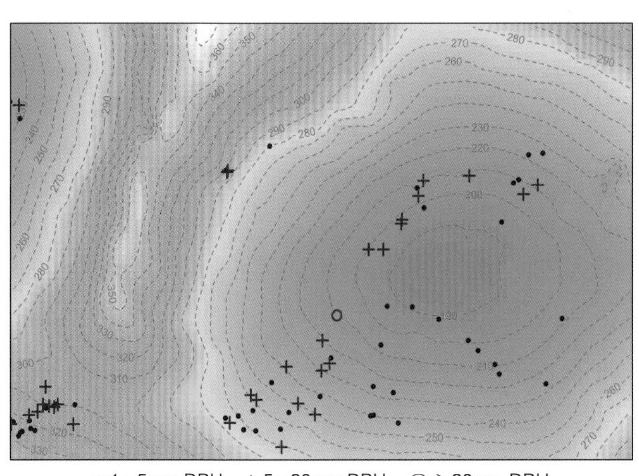

• 1~5 cm DBH　+ 5~20 cm DBH　○ ≥20 cm DBH
个体分布图 Distribution of individuals

径级分布表 DBH class

胸径区间 (Diameter class) (cm)	个体数 (No. of individuals in the plot)	比例 (Proportion) (%)
1~2	24	30.00
2~5	23	28.75
5~10	23	28.75
10~20	9	11.25
20~35	1	1.25
35~60	0	0.00
≥60	0	0.00

197 三脉叶荚蒾

sān mài yè jiā mí | Triplinervous Viburnum

Viburnum triplinerve Hand.-Mazz.
忍冬科 Caprifoliaceae

代码（SpCode）= VIBTRI
个体数（Individual number/15 hm^2）= 140
最大胸径（Max DBH）= 8.1 cm
重要值排序（Importance value rank）= 79

常绿灌木，高 2~4 m。全体无毛，枝褐色。叶常集生小枝顶，具 3 出脉，延伸至叶全长的约 3/4 处，近边缘前弯拱而相互网结。聚伞花序，顶生。果实近圆形，熟时紫褐色，有 1 条极细的浅腹沟。花期 4~5 月，果期 6~10 月。

Evergreen shrubs, 2–4 m tall. Glabrous, brak gray-brownish. Leaves often clustered at apices of branchlets, triplinerved, veins reaching 3/4 length of leaves, arched, anastomosing near margin. Inflorescence a compound umbel-like cyme, terminal. Fruit nearly rounded, purple-brownish, with 1 very small and shallow ventral groove. Fl. Apr.–May, fr. Jun.–Oct..

枝叶 Branch and leaves
摄影：王斌 Photo by：Wang Bin

叶背 Leaf back
摄影：王斌 Photo by：Wang Bin

果 Fruits
摄影：黄俞淞 Photo by：Huang Yusong

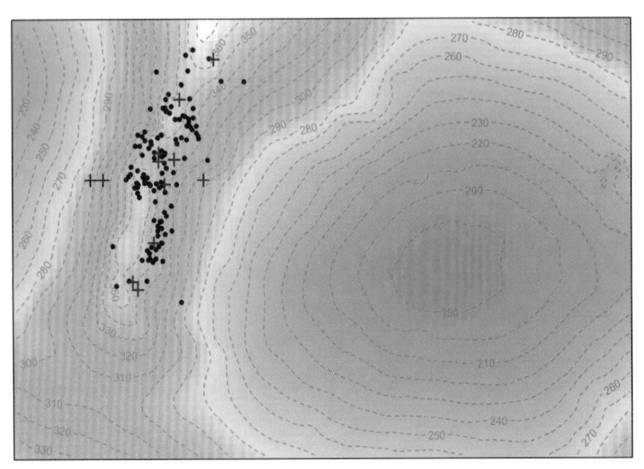

● 1~5 cm DBH + 5~20 cm DBH ○ ≥20 cm DBH
个体分布图 Distribution of individuals

径级分布表 DBH class

胸径区间 (Diameter class) (cm)	个体数 (No. of individuals in the plot)	比例 (Proportion) (%)
1~2	32	22.86
2~5	96	68.57
5~10	12	8.57
10~20	0	0.00
20~35	0	0.00
35~60	0	0.00
≥60	0	0.00

198 二歧破布木

èr qí pò bù mù | Furcate Cordia

Cordia furcans I. M. Johnst.
紫草科 Boraginaceae

代码（SpCode）= CORFUR
个体数（Individual number/15 hm^2）= 2
最大胸径（Max DBH）= 10.2 cm
重要值排序（Importance value rank）= 190

灌木或小乔木，高 5~10 m。树皮灰色。叶卵圆形或椭圆形，通常全缘，上面被硬毛，下面密生淡黄色短茸毛，稀近无毛。聚伞花序多花，顶生及侧生，侧生花序腋外生。核果红色或淡红色，椭圆形，被不规则浅裂的杯状宿萼承托。花期 11 月，翌年果期 1 月。

Shrubs or small trees, 5–10 m tall. Bark gray. Leaf blade ovate or elliptic, margin usually entire, adaxially hispid, abaxially densely yellowish pubescent, rarely subglabrous. Cymes terminal and lateral，lateral cymes extra-axillary. Drupes red or pale red, elliptic, with irregular lobed cupular persistent calyx. Fl. Nov., fr. Jan. of next year.

枝叶 Branch and leaves
摄影：黄俞淞 Photo by：Huang Yusong

叶 Leaves
摄影：黄俞淞 Photo by：Huang Yusong

花 Flowers
摄影：黄俞淞 Photo by：Huang Yusong

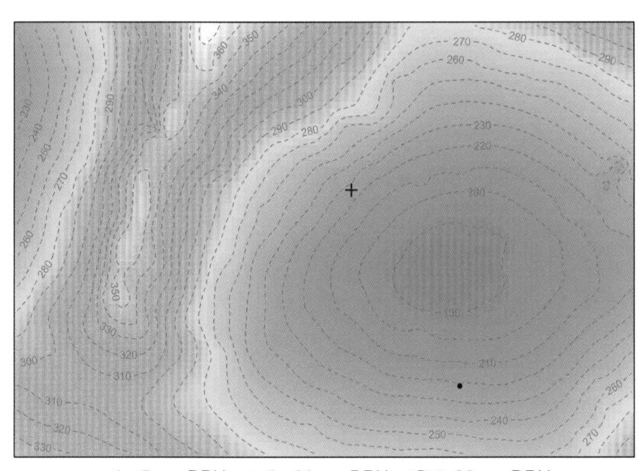

• 1~5 cm DBH　+ 5~20 cm DBH　○ ≥20 cm DBH
个体分布图 Distribution of individuals

径级分布表 DBH class

胸径区间 (Diameter class) (cm)	个体数 (No. of individuals in the plot)	比例 (Proportion) (%)
1~2	1	50.00
2~5	0	0.00
5~10	0	0.00
10~20	1	50.00
20~35	0	0.00
35~60	0	0.00
≥60	0	0.00

199 上思厚壳树

shàng sī hòu ké shù | Tsang's Ehretia

Ehretia tsangii Johnst.
紫草科 Boraginaceae

代码（SpCode）= EHRTSA
个体数（Individual number/15 hm^2）= 143
最大胸径（Max DBH）= 37.2 cm
重要值排序（Importance value rank）= 54

乔木或小乔木，高 3~10 m。枝灰褐色，无毛。叶椭圆形或长圆状椭圆形，全缘，无毛，下面仅脉腋间具柔毛。聚伞花序顶生及侧生，具短柔毛。核果黄色，直径约 5 mm，内果皮成熟时分裂为 4 个具单种子的分核。花期 3 月，果期 4 月。

Trees or small trees, 3–10 m tall. Branches gray-brown, glabrous. Leaf blade elliptic or oblong-elliptic, margin entire, glabrous, abaxially pubescent only in vein axils. Cymes terminal and lateral, pubescent. Drupes yellow, ca. 5 mm in diam., endocarp divided at maturity into 4 1-seeded pyrened. Fl. Mar., fr. Apr..

树干 Trunk
摄影：王斌 Photo by：Wang Bin

果 Fruit
摄影：黄俞淞 Photo by：Huang Yusong

花枝 Flowering branches
摄影：黄俞淞 Photo by：Huang Yusong

• 1~5 cm DBH + 5~20 cm DBH ○ ≥20 cm DBH
个体分布图 Distribution of individuals

径级分布表 DBH class

胸径区间 (Diameter class) (cm)	个体数 (No. of individuals in the plot)	比例 (Proportion) (%)
1~2	18	12.59
2~5	31	21.68
5~10	59	41.26
10~20	32	22.38
20~35	2	1.40
35~60	1	0.70
≥60	0	0.00

200 龙州厚壳树 lóng zhōu hòu ké shù | Longzhou Ehretia

Ehretia longzhouensis Y.S.Huang
紫草科 Boraginaceae

代码（SpCode）= EHRLON
个体数（Individual number/15 hm^2）= 2
最大胸径（Max DBH）= 3.2 cm
重要值排序（Importance value rank）= 187

小乔木，高 3~6 m。树皮灰褐色。叶长圆状披针形，长 5~10 cm，宽 2~4 cm, 基部楔形，全缘，侧脉 4~6 对。聚伞花序顶生或腋生，花冠筒钟状，白色，花柱无毛，柱头 2 裂。花期 7~8 月。

Small trees, 3–6 cm tall. Bark gray brown. Leaf blade oblong-elliptic, 5–10 cm long , and 2–4 cm wide, base cuneate, margin entire, lateral veins 4–6 pairs. Cymes terminal or axillary, corolla campanulate, white, style glabrous, stigma 2-lobed. Mature fruit not seen. Fl. Jul.–Aug..

花枝 Flowering branches
摄影：许为斌 Photo by：Xu Weibin

花 Flowers
摄影：许为斌 Photo by：Xu Weibin

叶 Leaves
摄影：许为斌 Photo by：Xu Weibin

• 1~5 cm DBH + 5~20 cm DBH ○ ⩾20 cm DBH
个体分布图 Distribution of individuals

径级分布表 DBH class

胸径区间 (Diameter class) (cm)	个体数 (No. of individuals in the plot)	比例 (Proportion) (%)
1~2	1	50.00
2~5	1	50.00
5~10	0	0.00
10~20	0	0.00
20~35	0	0.00
35~60	0	0.00
⩾60	0	0.00

201 西南猫尾木

xī nán māo wěi mù | Stipulate Dolichandrone

Dolichandrone stipulata (Wall.) Benth. et Hook. f.
紫葳科 Bignoniaceae

代码（SpCode）= DOLSTI
个体数（Individual number/15 hm^2）= 202
最大胸径（Max DBH）= 70.0 cm
重要值排序（Importance value rank）= 63

落叶乔木，高 10~15 m。嫩枝、嫩叶及花序轴密被黄褐色短柔毛。奇数羽状复叶长达 30 cm，小叶 7~11 枚，长椭圆形至椭圆状卵形，两面近无毛，有时两面被稀疏黑色腺点，全缘。花序为顶生总状聚伞花序，被锈黄色柔毛。蒴果披针形，长达 36 cm。花期 9~12 月，果期翌年 2~3 月。

Deciduous trees, 10–15 m tall. Twigs, tender leaf and rachis densely yellowish brown pubescent. Leaves odd-pinnate, up to 30 cm long, leaflets 7–11, oblong-elliptic to elliptic-ovate, both surfaces nearly glabrous, sometimes sparsely black gland spot, margin entire. Cymes terminal, rust yellow pubescent. Capsules lanceolate, up to 36 cm long. Fl. Sep.–Dec., fr. Feb.–Mar. of next year.

嫩叶 New Leaves
摄影：刘晟源 Photo by: Liu Shengyuan

枝 Branches
摄影：黄俞淞 Photo by: Huang Yusong

果枝 Fruiting branches
摄影：丁涛 Photo by: Ding Tao

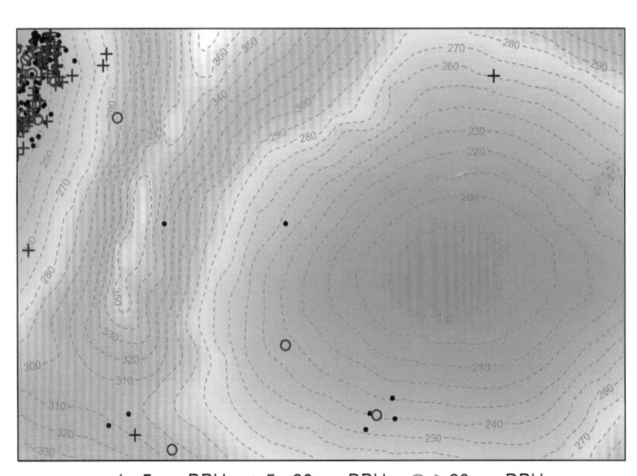

• 1~5 cm DBH + 5~20 cm DBH ○ ≥20 cm DBH
个体分布图 Distribution of individuals

径级分布表 DBH class

胸径区间 (Diameter class) (cm)	个体数 (No. of individuals in the plot)	比例 (Proportion) (%)
1~2	66	32.67
2~5	66	32.67
5~10	28	13.86
10~20	28	13.86
20~35	9	4.46
35~60	3	1.49
≥60	2	0.99

202 木蝴蝶

mù hú dié | Indian Trumpet Flower

Oroxylum indicum (Linn.) Benth. ex Kurz
紫葳科 Bignoniaceae

代码（SpCode）= OROIND
个体数（Individual number/15 hm^2）= 92
最大胸径（Max DBH）= 49.7 cm
重要值排序（Importance value rank）= 57

落叶小乔木，高 6~10 m。树皮灰褐色。奇数二回羽状复叶，对生，着生于茎干近顶端，小叶三角状卵形，基部近圆或心形，偏斜，叶片干后变蓝色。总状聚伞花序顶生，粗壮，花大，紫红色，花冠在傍晚开放，有恶臭气味。蒴果木质，果瓣具中肋，种子周翅薄如纸。花期 9~12 月。

Deciduous small trees, 6–10 m tall. Bark gray brown. Leaves 2-pinnately compound, opposite, borne nearly at stem apex, leaflets triangular-ovate, base subrounded or cordate, oblique, becoming blue after drying. Inflorescences terminal, strong, flowers purple red, usually open at night, with foul smell. Capsule woody, valves with midrib, seeds with papery wing. Fl. Sep.–Dec..

树干 Trunk
摄影：王斌 Photo by: Wang Bin

复叶 Compound leaf
摄影：王斌 Photo by: Wang Bin

花 Flower
摄影：黄俞淞 Photo by: Huang Yusong

• 1~5 cm DBH　+ 5~20 cm DBH　○ ≥20 cm DBH

个体分布图 Distribution of individuals

径级分布表 DBH class

胸径区间 (Diameter class) (cm)	个体数 (No. of individuals in the plot)	比例 (Proportion) (%)
1~2	3	3.26
2~5	4	4.35
5~10	6	6.52
10~20	46	50.00
20~35	31	33.70
35~60	2	2.17
≥60	0	0.00

203 美叶菜豆树

mĕi yè cài dòu shù | Fern-leaf China Doll Tree

Radermachera frondosa Chun et F. C. How
紫葳科 Bignoniaceae

代码（SpCode）= RADFRO
个体数（Individual number/15 hm^2）= 54
最大胸径（Max DBH）= 52.4 cm
重要值排序（Importance value rank）= 89

落叶乔木，高 7~20 m。小枝被微柔毛。二回羽状复叶，对生，小叶 5~7，椭圆形至卵形，纸质，顶端尾状渐尖。花序顶生，尖塔状，直立，长 30 cm，三歧分叉。花果期几乎全年。

Deciduous trees, 7–20 m tall. Branchlets puberulent. Leaves 2-pinnately compound, opposite, leaflets 5–7, elliptic to ovate, papery, apex caudate-acuminate. Inflorescences paniculate, terminal, erect, 30 cm long, 3-branched. Fl. and fr. almost seasons.

树干 Trunk
摄影：黄俞淞 Photo by：Huang Yusong

叶背 Leaf back
摄影：黄俞淞 Photo by：Huang Yusong

花序 Inflorescence
摄影：刘晟源 Photo by：Liu Shengyuan

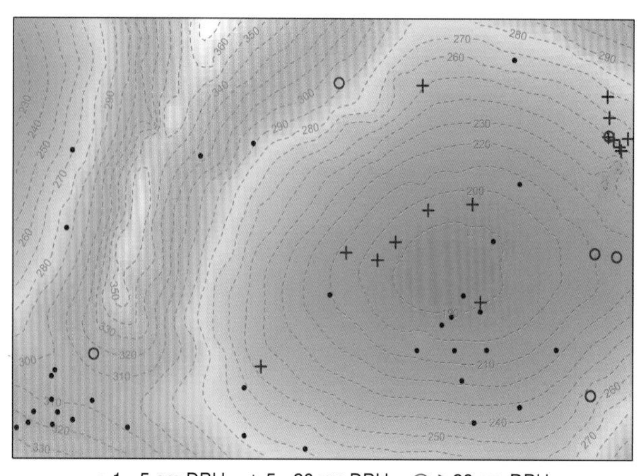

• 1~5 cm DBH + 5~20 cm DBH ○ ≥20 cm DBH
个体分布图 Distribution of individuals

径级分布表 DBH class

胸径区间 (Diameter class) (cm)	个体数 (No. of individuals in the plot)	比例 (Proportion) (%)
1~2	16	29.63
2~5	17	31.48
5~10	10	18.52
10~20	5	9.26
20~35	4	7.41
35~60	2	3.70
≥60	0	0.00

204 菜豆树

cài dòu shù | China Doll Tree

Radermachera sinica (Hance) Hemsl.
紫葳科 Bignoniaceae

代码（SpCode）= RADSIN
个体数（Individual number/15 hm^2）= 88
最大胸径（Max DBH）= 28.5 cm
重要值排序（Importance value rank）= 77

小乔木，高达 10 m。叶柄、叶轴、花序均无毛。二回羽状复叶，小叶卵形至卵状披针形，顶端尾状渐尖，基部阔楔形，全缘，两面均无毛。顶生圆锥花序，直立，长 25~35 cm。蒴果细长，下垂，长达 85 cm。种子椭圆形，连翅长约 2 cm。花期 5~9 月，果期 10~12 月。

Small trees, up to 10 m tall. Petioles, leaf axis, and inflorescences glabrous. Leaves 2 pinnately compound, leaflets ovate-lanceolate, apex caudate-acuminate, base broadly cuneate, entire, both surfaces glabrous. Inflorescences paniculate, terminal, erect, 25–35 cm long. Capsule slender, nodding, ca. 85 cm long, seeds elliptic, including wing ca. 2 cm long. Fl. May–Sep., fr. Oct.–Dec..

树干 Trunk
摄影：王斌 Photo by：Wang Bin

花和复叶 Flower and compound leaves
摄影：徐永福 Photo by：Xu Yongfu

果和花 Fruit and flower
摄影：徐永福 Photo by：Xu Yongfu

• 1~5 cm DBH + 5~20 cm DBH ○ ≥20 cm DBH

个体分布图 Distribution of individuals

径级分布表 DBH class

胸径区间 (Diameter class) (cm)	个体数 (No. of individuals in the plot)	比例 (Proportion) (%)
1~2	18	20.45
2~5	23	26.14
5~10	17	19.32
10~20	15	17.05
20~35	15	17.05
35~60	0	0.00
≥60	0	0.00

205 羽叶楸

yǔ yè qiū | Yellow Snake Tree

Stereospermum colais (Buch. -Ham. ex Dillwyn) Mabberley
紫葳科 Bignoniaceae

代码（SpCode）= STECOL
个体数（Individual number/15 hm^2）= 207
最大胸径（Max DBH）= 45.9 cm
重要值排序（Importance value rank）= 45

落叶乔木，高 15~20(35)m。一回羽状复叶，长 25~50 cm，小叶 3~6 对，长椭圆形，基部阔楔形至圆形，全缘，无毛。圆锥花序顶生，长 20~40 cm。蒴果细长，四棱柱形，微弯曲，长 30~70 cm，粗约 1 cm。种子卵圆形，两端具有白色膜质翅。花期 5~7 月，果期 9~I1 月。

Deciduous trees, 15–20(30) m tall. Leaves 1-pinnately, 25–50 cm long, leaflets 3–6 pairs, oblong-elliptic, base broadly cuniform to rounded, entire, glabrous. Inflorescences paniculate, terminal, 20–40 cm long. Capsule slender, 4-angular, slightly curved, 30–70 cm long, ca. 1 cm in diam., seeds ovate-rounded, white membranous winged at apices. Fl. May–Jul., fr. Sep.–Nov..

树干 Trunk
摄影：王斌 Photo by：Wang Bin

花枝 Flowering branches
摄影：黄俞淞 Photo by：Huang Yusong

复叶 Compound leaves
摄影：黄俞淞 Photo by：Huang Yusong

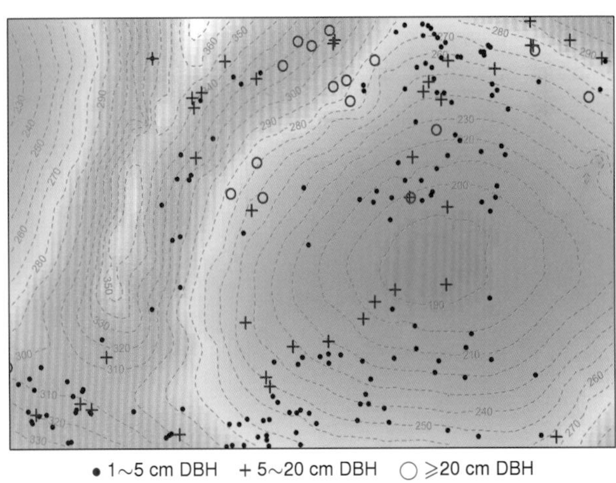
• 1~5 cm DBH + 5~20 cm DBH ○ ≥20 cm DBH
个体分布图 Distribution of individuals

径级分布表 DBH class

胸径区间 (Diameter class) (cm)	个体数 (No. of individuals in the plot)	比例 (Proportion) (%)
1~2	94	45.41
2~5	59	28.50
5~10	21	10.14
10~20	17	8.21
20~35	13	6.28
35~60	3	1.45
≥60	0	0.00

206 南川紫珠 nán chuān zǐ zhū | Rosthorn's Beautyberry

Callicarpa bodinieri var. *rosthornii* (Diels) Rehd.
马鞭草科 Verbenaceae

代码（SpCode）= CALBOD
个体数（Individual number/15 hm^2）= 1
最大胸径（Max DBH）= 2.7 cm
重要值排序（Importance value rank）= 208

灌木，高 1~3 m。叶膜质，倒披针形或倒卵状长圆形，两面具密集的腺点，下面具灰色星状毛，基部下延成狭楔形，顶端渐尖，上半部有锯齿。花期 6~7 月，果期 8~11 月。

Shrubs, 1–3 m tall. Leaf blade membranous, oblanceolate to obovate-oblong, densely glandular, abaxially grayish stellate pubescent, base narrowly cuneate, apex acuminate, margin serrate above middle. Fl. Jun.–Jul., fr. Aug.–Nov..

枝叶 Branches and leaves
摄影：黄俞淞 Photo by：Huang Yusong

果序 Infructescence
摄影：黄俞淞 Photo by：Huang Yusong

叶 Leaves
摄影：黄俞淞 Photo by：Huang Yusong

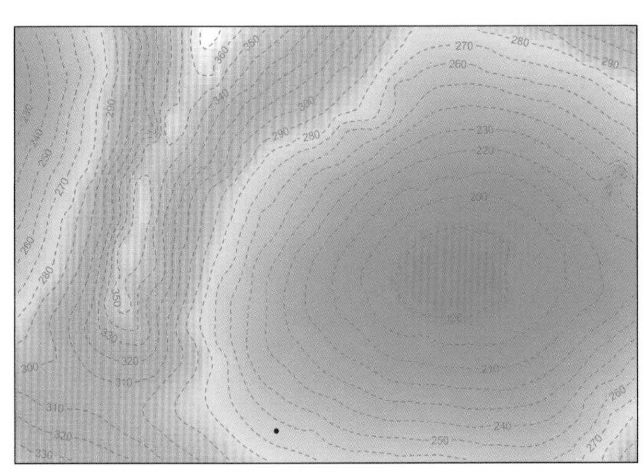

• 1~5 cm DBH + 5~20 cm DBH ○ ≥20 cm DBH
个体分布图 Distribution of individuals

径级分布表 DBH class

胸径区间 (Diameter class) (cm)	个体数 (No. of individuals in the plot)	比例 (Proportion) (%)
1~2	0	0.00
2~5	1	100.00
5~10	0	0.00
10~20	0	0.00
20~35	0	0.00
35~60	0	0.00
≥60	0	0.00

207 华紫珠

huá zǐ zhū | Chinese Beautyberry

Callicarpa cathayana H. T. Chang
马鞭草科 Verbenaceae

代码（SpCode）= CALCAT
个体数（Individual number/15 hm^2）= 1
最大胸径（Max DBH）= 2.0 cm
重要值排序（Importance value rank）= 212

灌木，高 1.5~3 m。幼嫩稍有星状毛，老后脱落。叶片椭圆形或卵形，顶端渐尖，基部楔形，两面近于无毛，而有显著的红色腺点，边缘密生细锯齿。聚伞花序 3~4 次分歧，略有星状毛。果实球形，紫色，径约 2 mm。花期 5~7 月，果期 8~11 月。

Shrubs, 1.5–3 m tall. Young branchlets slightly stellate tomentose, glabrescent. Leaf blade elliptic or ovate, apex acuminate, base cuneate, both surfaces nearly glabrous, with conspicuous red gland, margin densely serrulate. Cymes 3-4-branched, slightly stellate. Fruit globose, red, ca. 2 mm in diam.. Fl. May–Jul., fr. Aug.–Nov..

果枝 Fruiting branches
摄影：刘晟源 Photo by：Liu Shengyuan

叶 Leaf
摄影：刘晟源 Photo by：Liu Shengyuan

果序 Infructescence
摄影：刘晟源 Photo by：Liu Shengyuan

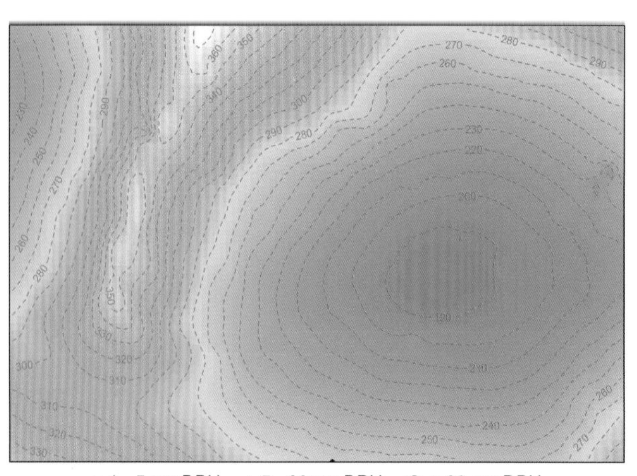

• 1~5 cm DBH + 5~20 cm DBH ○ ≥20 cm DBH
个体分布图 Distribution of individuals

径级分布表 DBH class

胸径区间 (Diameter class) (cm)	个体数 (No. of individuals in the plot)	比例 (Proportion) (%)
1~2	0	0.00
2~5	1	100.00
5~10	0	0.00
10~20	0	0.00
20~35	0	0.00
35~60	0	0.00
≥60	0	0.00

208 白毛长叶紫珠 bái máo cháng yè zǐ zhū | Floccose Beautyberry

Callicarpa longifolia var. *floccosa* Schauer
马鞭草科 Verbenaceae

代码（SpCode）= CALLON
个体数（Individual number/15 hm^2）= 20
最大胸径（Max DBH）= 9.5 cm
重要值排序（Importance value rank）= 136

落叶灌木，高 2-5 m。小枝稍四棱形，密被粉屑状灰白色星状毛。叶片长圆状椭圆形，叶背密被粉屑状灰白色星状毛。聚伞花序，花序和花的各部分均密被粉屑状灰白色星状毛。果实圆形。花期 8 月，果期 9~11 月。

Deciduous shrubs, 2–5 m tall. Branchlets slightly 4-angles, densely gray mealy stellate hairs. Leaf blade oblong-elliptic, abaxially densely gray mealy stellate hairs. Cymes, Inflorescence and flowers densely gray mealy stellate hairs. Fruit rounded. Fl. Aug., fr. Sep.–Nov..

花枝 Flowering branch
摄影：黄俞淞 Photo by：Huang Yusong

枝叶 Branch and leaves
摄影：黄俞淞 Photo by：Huang Yusong

果 Fruits
摄影：黄俞淞 Photo by：Huang Yusong

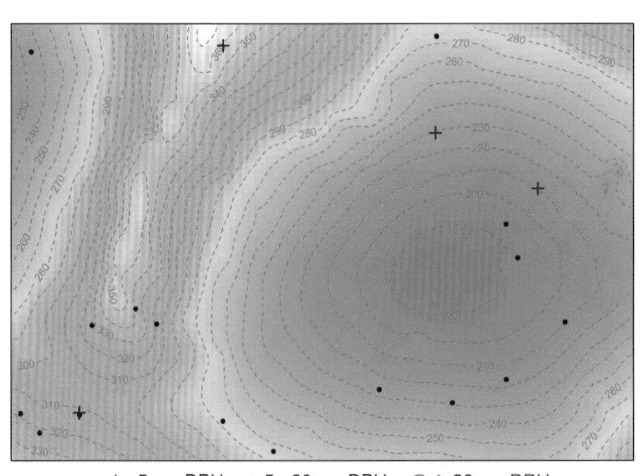

• 1~5 cm DBH + 5~20 cm DBH ○ ≥20 cm DBH
个体分布图 Distribution of individuals

径级分布表 DBH class

胸径区间 (Diameter class) (cm)	个体数 (No. of individuals in the plot)	比例 (Proportion) (%)
1~2	6	30.00
2~5	10	50.00
5~10	4	20.00
10~20	0	0.00
20~35	0	0.00
35~60	0	0.00
≥60	0	0.00

209 三对节

sān duì jié | Serrate Glorybower

Clerodendrum serratum (Linn.) Moon
马鞭草科 Verbenaceae

代码（SpCode）= CLESER
个体数（Individual number/15 hm^2）= 1
最大胸径（Max DBH）= 1.8 cm
重要值排序（Importance value rank）= 213

灌木，高 1~4 m。小枝四棱形或略呈四棱形。叶片厚纸质，对生或三叶轮生，倒卵状长圆形或长椭圆形，边缘具锯齿，两面疏生短柔毛，叶柄长 0.5~1 cm 或近无柄。聚伞花序组成直立、开展的圆锥花序，顶生。核果近球形，绿色，后转黑色。花果期 6~12 月。

Shrubs, 1–4 m tall. Branchlets 4-angled or slightly 4-angled. Leaf blade thickly papery, opposite or in three, obovate-oblong or oblong-elliptic, margin serrulate, both surfaces thinly pubescent. Petiole 0.5–1 cm long or leaf subsessile. Inflorescences terminal thyrses, erect. Drupes subglobose, green when, becoming black. Fl. and fr. Jun.–Dec..

花枝　Flowering branch
摄影：黄俞淞　Photo by: Huang Yusong

果序　Infructescence
摄影：黄俞淞　Photo by: Huang Yusong

枝叶　Branch and leaves
摄影：黄俞淞　Photo by: Huang Yusong

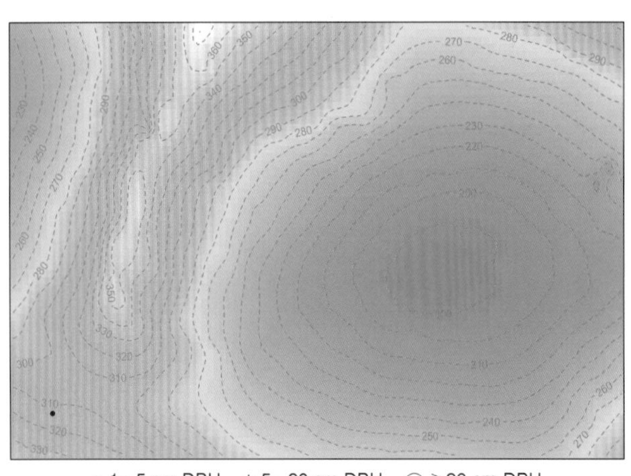

• 1~5 cm DBH　+ 5~20 cm DBH　○ ≥20 cm DBH
个体分布图　Distribution of individuals

径级分布表　DBH class

胸径区间 (Diameter class) (cm)	个体数 (No. of individuals in the plot)	比例 (Proportion) (%)
1~2	1	100.00
2~5	0	0.00
5~10	0	0.00
10~20	0	0.00
20~35	0	0.00
35~60	0	0.00
≥60	0	0.00

210 垂茉莉

chuí mò lì | Nodding Clerodendron

Clerodendrum wallichii Merr.
马鞭草科 Verbenaceae

代码（SpCode）= CLEWAL
个体数（Individual number/15 hm^2）= 934
最大胸径（Max DBH）= 8.6 cm
重要值排序（Importance value rank）= 27

常绿直立灌木或小乔木，高 2~4 m。小枝锐四棱形或呈翅状。叶片长圆形或长圆状披针形。聚伞花序排列成圆锥状，下垂，每聚伞花序对生或交互对生。核果球形，初时黄绿色，成熟后紫黑色。花果期 10 月至翌年 4 月。

Evergreen shrubs or small trees, 2–4 m tall. Branchlets 4-angled, winged. Leaf blade oblong ro oblong-lanceolate. Thyrses pendent, each cyme opposite or alternately opposite. Drupe globose, yellowish green when young, atropurpureus when mature, Fl. and fr. Oct.–Apr. of next year.

花序 Inflorescence
摄影：王斌 Photo by: Wang Bin

叶 Leaves
摄影：黄俞淞 Photo by: Huang Yusong

果枝 Fruiting branch
摄影：黄俞淞 Photo by: Huang Yusong

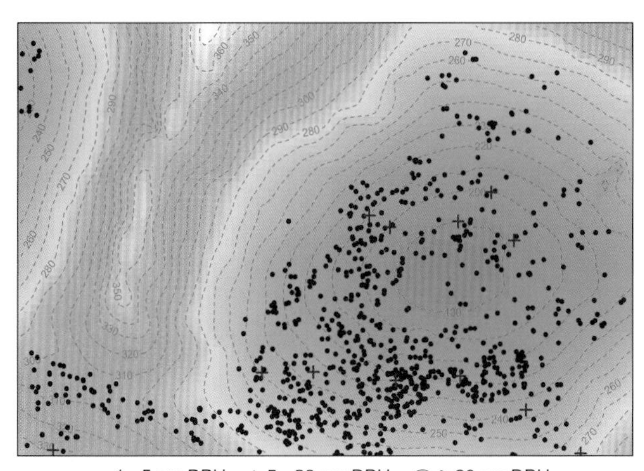

• 1~5 cm DBH + 5~20 cm DBH ○ ≥20 cm DBH
个体分布图 Distribution of individuals

径级分布表 DBH class

胸径区间 (Diameter class) (cm)	个体数 (No. of individuals in the plot)	比例 (Proportion) (%)
1~2	465	49.79
2~5	457	48.93
5~10	12	1.28
10~20	0	0.00
20~35	0	0.00
35~60	0	0.00
≥60	0	0.00

211 滇桂豆腐柴

diān guì dòu fū chái | Adjoining Premna

Premna confinis Pei et S. L. Chen ex C. Y. Wu
马鞭草科 Verbenaceae

代码（SpCode）= PRECON
个体数（Individual number/15 hm^2）= 21
最大胸径（Max DBH）= 6.2 cm
重要值排序（Importance value rank）= 153

常绿灌木至小乔木，高 1.5~6 m。小枝密被糠粃状腺点，无毛。叶片革质，长圆形至披针形，稀椭圆形，全缘或波状，无毛，背面密被暗黄色腺点，叶柄无毛，密被腺点。圆锥花序顶生。果实紫红色，有腺点，微有瘤状突起。花期 5 月。

Evergreen shrubs to small trees, 1.5–6 m tall. Branchlets densely squarrose glandular spots, glabrous. Leaf bladeleathery, oblong to lanceolate, sometimes elliptic, entire or undate, glabrous, abaxial densely dark-yellow glandular spots, petiole glabrous, densely glandular spots. Panicle terminal. Fruit purpl-red, slightly tubercular, glandular. Fl. May.

花枝 Flowering branch
摄影：黄俞淞 Photo by：Huang Yusong

花 Flowers
摄影：黄俞淞 Photo by：Huang Yusong

果枝 Fruiting branch
摄影：黄俞淞 Photo by：Huang Yusong

• 1~5 cm DBH + 5~20 cm DBH ○ ≥20 cm DBH
个体分布图 Distribution of individuals

径级分布表 DBH class

胸径区间 (Diameter class) (cm)	个体数 (No. of individuals in the plot)	比例 (Proportion) (%)
1~2	11	52.38
2~5	9	42.86
5~10	1	4.76
10~20	0	0.00
20~35	0	0.00
35~60	0	0.00
≥60	0	0.00

212 黄毛豆腐柴

huáng máo dòu fū chái | Yellow-hairy Premna

Premna fulva Craib
马鞭草科 Verbenaceae

代码（SpCode）= PREFUL
个体数（Individual number/15 hm^2）= 14
最大胸径（Max DBH）= 5.2 cm
重要值排序（Importance value rank）= 155

灌木或小乔木，有时攀援状。幼枝密被黄色平展长柔毛，老枝变无毛且转红褐色。叶片纸质，卵圆形、长圆状倒卵圆形或椭圆形等，叶背密被柔毛似毡。聚伞花序伞房状，顶生。核果卵形至球形，成熟时黑色，有瘤突。花果期 5~10 月。

Shrubs or trees, sometimes climbing. Branchlets yellow villous when young, glabrous when old, red-brown. Leaf brade papery, ovate, ovate-lanceolate, elliptic, oblong-ovate or subrounded, abaxial densely tomentose. Cyme corymbose, terminal. Drupe ovate to globose, black when mature, tuberculate. Fl. and fr. May–Oct..

花枝 Flowering branch
摄影：黄俞淞 Photo by：Huang Yusong

花序 Inflorescence
摄影：丁涛 Photo by：Ding Tao

果序 Infructescence
摄影：刘晟源 Photo by：Liu Shengyuan

• 1~5 cm DBH　+ 5~20 cm DBH　○ ≥20 cm DBH
个体分布图 Distribution of individuals

径级分布表 DBH class

胸径区间 (Diameter class) (cm)	个体数 (No. of individuals in the plot)	比例 (Proportion) (%)
1~2	7	50.00
2~5	6	42.86
5~10	1	7.14
10~20	0	0.00
20~35	0	0.00
35~60	0	0.00
≥60	0	0.00

213 广西牡荆

guǎng xī mǔ jīng | Guangxi Chastetree

Vitex kwangsiensis C. Pei
马鞭草科 Verbenaceae

代码（SpCode）= VITKWA
个体数（Individual number/15 hm^2）= 2470
最大胸径（Max DBH）= 46.0 cm
重要值排序（Importance value rank）= 3

落叶乔木。树皮常灰白色。复叶具 2~5 小叶，通常为 3，对生，小叶卵状披针形至卵形，全缘或具齿，顶端渐尖或尾状尖，叶背有金黄色腺点，两侧小叶较小。花序梗近无毛，花冠橙黄色，外面有细毛。花期 5~6 月，果期 7~9 月。

Diciduous trees. Bark usually grayish white. Leaves 2–5-foliolate, usually 3, opposite, leaflets ovate-lanceolate to ovate, entire or toothed, apex acuminate to caudate-acuminate, abaxial with golden yellow. Fl. May–Jun., fr. Jul.–Sep..

树干 Trunk
摄影：王斌 Photo by：Wang Bin

嫩枝 New branch and leaves
摄影：王斌 Photo by：Wang Bin

果枝 Fruiting branch
摄影：黄俞淞 Photo by：Huang Yusong

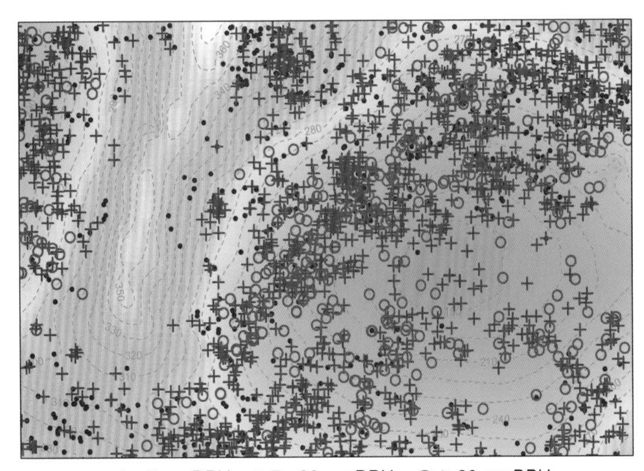

• 1~5 cm DBH + 5~20 cm DBH ○ ⩾20 cm DBH
个体分布图 Distribution of individuals

径级分布表 DBH class

胸径区间 (Diameter class) (cm)	个体数 (No. of individuals in the plot)	比例 (Proportion) (%)
1~2	264	10.69
2~5	413	16.72
5~10	513	20.77
10~20	926	37.49
20~35	337	13.64
35~60	17	0.69
⩾60	0	0.00

214 山牡荆

shān mǔ jīng | Five-leaf Chastetree

Vitex quinata (Lour.) F. N. Williams
马鞭草科 Verbenaceae

代码（SpCode）= VITQUI
个体数（Individual number/15 hm^2）= 187
最大胸径（Max DBH）= 10.2 cm
重要值排序（Importance value rank）= 75

常绿灌木或乔木，高 4~12 m。树皮灰褐色至深褐色。掌状复叶，对生，有 3~5 小叶，小叶片倒卵状椭圆形至倒卵形，基部楔形至阔楔形，叶背具金黄色腺点。圆锥花序式顶生。核果球形或倒卵形，成熟后呈黑色。花期 5~7 月，果期 8~9 月。

Evergreen shrubs or trees, 4–12 m tall. Bark grayish brown. Lesves 3–5-foliolate, opposite, leaflets obovate-elliptic to obovate, base cuneate to wide cuneate, leaf abaxial often with golden-yellow glandular punctate. Panicles terminal. Drupe globose or obovoid, black at maturity. Fl. May–Jul., fr. Aug.–Sep..

植株 Whole plant
摄影：黄俞淞 Photo by：Huang Yusong

花序 Inflorescence
摄影：黄俞淞 Photo by：Huang Yusong

复叶 Compound leaves
摄影：王斌 Photo by：Wang Bin

• 1~5 cm DBH + 5~20 cm DBH ○ ≥20 cm DBH
个体分布图 Distribution of individuals

径级分布表 DBH class

胸径区间 (Diameter class) (cm)	个体数 (No. of individuals in the plot)	比例 (Proportion) (%)
1~2	110	58.82
2~5	67	35.83
5~10	9	4.81
10~20	1	0.53
20~35	0	0.00
35~60	0	0.00
≥60	0	0.00

215 弄岗假糙苏

nòng gǎng jiǎ cāo sū | Longgang Paraphlomis

Paraphlomis longgangensis Y.S.Huang
唇形科 Lamiaceae

代码（SpCode）= PARLON
个体数（Individual number/15 hm^2）= 1
最大胸径（Max DBH）= 1.7 cm
重要值排序（Importance value rank）= 214

草本或亚灌木。茎单生，四棱形，具槽，高达 2.5 m。叶椭圆形或椭圆状卵形，长 10~30 cm，宽 5~10 cm，基部圆钝或微心形，边缘具小突尖的锯齿，叶柄长 8~16 cm。轮伞花序多花，花萼紫红色，冠檐二唇形，上唇淡黄色，下唇紫红色。小坚果黑色。花期 7~8 月，果期 9~10 月。

Herbs or subshrubs. Stems solitary, tetragonous, grooved, up to 2.5 m tall. Leaves elliptic or elliptic-ovate, 10–30 cm long, and 5–10 cm wide, base rounded or slightly cordate, margin conspicuously serrate, petiole 8–16 cm long. Verticillasters many flowered, calyx red, upper lip faint yellow, lower lip red. Nutlers black. Fl. Jul.–Aug., fr. Sep.–Oct..

植株 Whole plant
摄影：黄俞淞 Photo by：Huang Yusong

叶 Leaves
摄影：黄俞淞 Photo by：Huang Yusong

花 Flowers
摄影：黄俞淞 Photo by：Huang Yusong

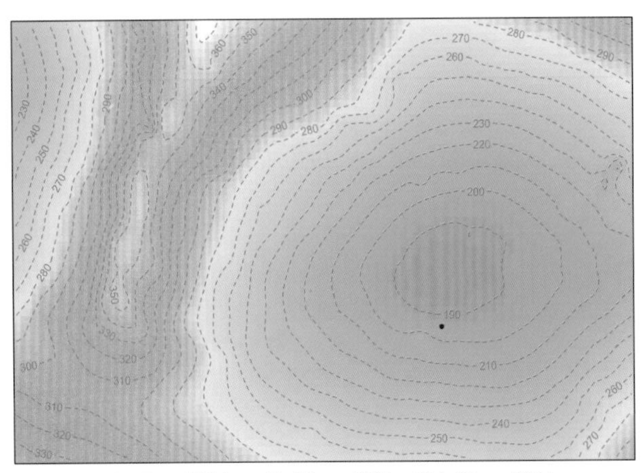

个体分布图 Distribution of individuals

径级分布表 DBH class

胸径区间 (Diameter class) (cm)	个体数 (No. of individuals in the plot)	比例 (Proportion) (%)
1~2	1	100.00
2~5	0	0.00
5~10	0	0.00
10~20	0	0.00
20~35	0	0.00
35~60	0	0.00
≥60	0	0.00

216 剑叶龙血树

jiàn yè lóng xuě shù | Sword-leaf dracaena

Dracaena cochinchinensis (Lour.) S. C. Chen
龙舌兰科 Agavaceae

代码（SpCode）= DRACOC
个体数（Individual number/15 hm^2）= 358
最大胸径（Max DBH）= 21.4 cm
重要值排序（Importance value rank）= 34

常绿灌木或小乔木，高可达 5~15 m。树皮灰白色，老时灰褐色，光滑。叶聚生在茎、分枝或小枝顶端，剑形，叶基部抱茎。圆锥花序长 40 cm 以上，花每 2~5 朵簇生。浆果直径约 8~12 mm，橘黄色。花期 3 月，果期 7~8 月。

Evergreen shrubs or treelike, up to 5–15 m tall. Bark grayish white, becoming grayish brown with age, smooth. Leaves crowed at apex of branches, sword-shaped, base completely covering internode. Panicle terminal, more than 40 cm long, flowers in clusters of 2–5; Berry orange, 8–12 mm in diam. Fl. Mar., fr. Jul.–Aug..

树干　Trunk
摄影：黄俞淞　Photo by：Huang Yusong

植株　Whole plants
摄影：王斌　Photo by：Wang Bin

花序　Inflorescence
摄影：黄俞淞　Photo by：Huang Yusong

• 1~5 cm DBH　+ 5~20 cm DBH　○ ≥20 cm DBH
个体分布图　Distribution of individuals

径级分布表　DBH class

胸径区间 (Diameter class) (cm)	个体数 (No. of individuals in the plot)	比例 (Proportion) (%)
1~2	0	0.00
2~5	144	40.22
5~10	207	57.82
10~20	6	1.68
20~35	1	0.28
35~60	0	0.00
≥60	0	0.00

217 董棕

dǒng zōng | Giant Fishtail-palm

Caryota obtusa Griff.

棕榈科 Arecaceae

代码（SpCode）= CAROBT

个体数（Individual number/15 hm^2）= 61

最大胸径（Max DBH）= 52.5 cm

重要值排序（Importance value rank）= 39

乔木状，高 5~25 m。茎直立，黑褐色，具明显的杯状叶痕。叶弓状下弯，二级羽片宽楔形或狭的斜楔形，边缘具规则的齿缺。佛焰苞长 30~45 cm，花序具多数、密集的穗状分枝花序。果实球形至扁球形，成熟时红色。花期 6~10 月，果期 5~10 月。

Treelike, 5–25 m tall. Stems solitary, black brown, conspicuous cupulate leaf scar; secondary pinnas wide cuneate or narrow oblique cuneate, with regular jagged margins. Spathe 30–45 cm, inflorescence with many and dense spicate branches. Fruit globose or oblate, red at maturity. Fl. Jun.–Oct., fr. May–Oct..

树干 Trunk

摄影：王斌 Photo by: Wang Bin

果序 Infructescences

摄影：刘晟源 Photo by: Liu Shengyuan

植株 Whole plants

摄影：刘晟源 Photo by: Liu Shengyuan

• 1~5 cm DBH + 5~20 cm DBH ○ ≥20 cm DBH

个体分布图 Distribution of individuals

径级分布表 DBH class

胸径区间 (Diameter class) (cm)	个体数 (No. of individuals in the plot)	比例 (Proportion) (%)
1~2	0	0.00
2~5	5	8.20
5~10	1	1.64
10~20	11	18.03
20~35	21	34.43
35~60	23	37.70
≥60	0	0.00

致　谢

在群峰耸立、壁立千仞的北热带喀斯特峰丛山区，建设投影面积为15hm^{2}的喀斯特原生性森林大型监测样地，是一个前所未有的巨大挑战。

弄岗样地的建设者们于2008年开始在中越边境的原生性喀斯特季节性雨林中进行了多次勘察选址，最终决定在广西弄岗国家级自然保护区的弄岗片森林内建立长期监测样地，2010年初完成样地的标定和测绘，2011年全面完成首次植被调查，期间投入的人力累计达到2500多个工作日。

在整个样地建设过程中得到了各方面的大力支持与帮助，令我们得以克服常人无法想象的困难和艰辛。藉此样地手册出版之际，诚挚地感谢有关单位各级领导、专家和同事为样地建设作出的难以磨没的贡献：

(1) 中国科学院植物研究所马克平研究员的支持与鼓励对样地建设起到举足轻重的作用；中国科学院植物研究所郭柯研究员、米湘成副研究员、任海保博士和赖江山博士等参加了样地选址并全程给予技术指导；

(2) 广西弄岗国家级自然保护区管理局蒙渊君、游志峰、唐华兴、梁海峰、陈天波、农正权等对样地建设提供诸多帮助，其宝贵的基础资料和野外调查的丰富经验使我们胸有成竹，在调查队的后勤保障方面也几乎有求必应，使野外调查顺利开展；

(3) 广西植物研究所领导对野外监测平台的建设十分重视并提供了诸多方便，叶铎、何兰军、卢清柏、王新桂、吕仕洪、吴望辉、张德楠、潘复静、韦春强、刘明超、张中峰、徐广平、张建亮、钟军弟、尤业明、王三秋、黎彦余、王静、罗莉等科技人员和研究生参加野外调查；文淑均、郭屹立、黄甫昭、白坤栋、何运林、蒋裕良等参与了样品采集及数据分析；李卫华、唐海平两位师傅为调查队做了细

心周到的后勤服务；

(4) 广西师范大学生命科学学院为样地建设投入了大量的人力，梁士楚教授对野外调查工作给予了大力支持；广西师范大学2007级生态学本科生，罗亚进、刘存燕、黄传昌、王书明、陆振诚、唐君萍、吴孟益、张秀珍、李东梅、廖秋敏、杨丽平、林钰、陈映、梁锋、覃月娥、廖继肖、黄莹、何健津、廖多网、许承德等参与了植被调查；2008级、2009级、2010级共20名本科实习生参加了后续辅助样地和卫星样地调查；

(5) 弄岗保护站护林员苏理民、赵贵民、黄海龙、农伟宏、梁立新、苏海生，陇垒屯和坡那屯村民的苏理生、黄春、苏金辉、赵艳辉、苏江龙、何汉飞、黄爱明、闭兰川、梁爱芳等参与样地建设及后勤工作。

本书由中国科学院生物多样性委员会组织并资助，调查工作得到国家科技支撑计划课题“重要森林物种资源监测技术与示范”（2008BAC39B02）、广西“新世纪十百千人才工程”专项（2007219）和广西科技计划（桂科攻1598016-10）等项目支持。

承广西龙州县文联主席严造新先生惠赐摄影佳作，令本书增色不少；上海辰山植物园刘夙博士为本书翻译拟定植物英文名；中山大学逸仙学院博雅教育计划2015级李博伦同学认真校对书稿，使本书避免一些错误；中国林业出版社于界芬女士及其同事对本书的出版辛勤付出；谨此一并致以谢忱。

编者

2016年10月

Acknowledgements

It was an unprecedented great challenge to establish a large primitive karst forest dynamics plot with 15 hm^2 projected area in the northern tropical karst peak-cluster mountain area, where the mountain peaks tower into the sky and the precipices stand bolt upright.

The Nonggang plot builders conducted numbers of surveys and site selections in the primitive karst seasonal rainforest in the China-Vietnam boarder since 2008, and decided to establish a long-term monitoring forest plot in the Nongang part of Nonggang National Nature Reserve. The calibration and mapping of the plot was completed at the beginning of 2010. The first investigation of vegetation was fully completed in 2011. It costed more than 2500 working days in the construction period.

Many people and organizations gave us strong supports and helps for our plot establishment. They helped us to overcome the formidable difficulties and hardships. Therefore, we would like to take this opportunity of this book publication to greatly acknowledge all the leaders, experts and colleagues at all levels of relevant units for their indelible contributions to the plot construction:

(1) The support and encouragement from Professor Keping Ma of the Institute of Botany, Chinese Academy of Sciences played an important role in the plot construction. Professor Ke Guo, Associate Professor Xiangcheng Mi, Drs. Haibao Ren and Jiangshan Lai of the Institute of Botany, Chinese Academy of Sciences participated in the plot site selection and gave the technical guidance during the whole process.

(2) Yuanjun Meng, Zhifeng You, Huaxing Tang, Haifeng Liang, Tianbo Chen and Zhengquan Nong in Guangxi Nonggang National Nature Reserve Administration provided a lot of helps for the plot construction. Their valuable basic information and extensive field survey experience made us confident. They satisfied almost all of our logistical requirements, and enabled the well development of field survey.

(3) The leaders of Guangxi Institute of Botany attached great importance to the construction of the field monitoring platform and provided a lot of convenience. Many scientific and technical personnels and graduate students such as Duo Ye, Lanjun He, Qingbai Lu, Xingui Wang, Shihong Lv, Wanghui Wu, Denan Zhang, Fujing Pan, Chunqiang Wei, Mingchao Liu, Zhongfeng Zhang, Guangping Xu, Jianliang Zhang, Jundi Zhou,Yeming You, Sanqiu Wang, Yanyu

Li, Jing Wang and Li Luo participated in the field investigation. Shujun Wen, Yili Guo, Fuzhao Huang, Kundong Bai, Yunlin He and Yuliang Jiang participated in sample collection and data analysis. And Weihua Li and Haiping Tang made careful service to the investigation team.

(4) The College of Life Sciences of Guangxi Normal University pumped a lot of humanpower, and Professor Shichu Liang gave strong support to the field investigation work. The undergraduates majoring in ecology of 2007 in Guangxi Normal University, Yajin Luo, Cunyan Liu, Chuanchang Huang, Shuming Wang, Zhencheng Lu, Junping Tang, Feng Liang, Mengyi Wu, Yuee Qin, Xiuzhen Zhang, Dongmei Li, Liping Yang, Yu Lin, Jixiao Liao, Ying Huang, Ying Chen, Qiumin Liao, Jianjin He, Duowang Liao and Chengde Xu participated in the investigation of vegetation. And a total of 20 undergraduate students of 2008, 2009 and 2010 participated in the investigation of following auxiliary and satellite plots.

(5) The forest rangers of Nonggang Protection Station such as Limin Su, Guimin Zhao, Hailong Huang, Lixin Liang, and Haisheng Su, the villagers of Longleitun and Ponatun such as Lisheng Su, Chun Huang, Jinhui Su, Yanhui Zhao, Jianglong Su, Hanfei He, Aiming Huang and Aifang Liang participated in sample construction and logistic work.

This book is organized and funded by the Biodiversity Committee of Chinese Academy of Sciences. The survey got the financially supported by the National Key Technology Research and Development Program of China "Monitoring Technology and Demonstration of Important Forest Species Resources" (2008BAC39B02), Guangxi Special Fund "National Tens-Hundreds-Thousands Talent Introducing Program in the New Centuary" (2007219), and Guangxi Plan of Science and Technology (1598016-10).

The photographs from Zaoxin Yan, the chairman of The Art and Literary Association of Longzhou County, make the book more exciting. Doctor Su Liu of Shanghai Chenshan Botanical Gardern translated and determined the plant English names for this book. Bolun Li, 2015 undergraduate of Liberal Learning Program, Yat-sen School, Sun Yat-sen University, carefully proofread the manuscript, made the book avoid some mistakes; Ms. Jiefen Yu and her colleagues of China Forestry Publishing House, have done hard works on the publication of this book. We would like to extend our sincere gratitude to all these people.

Authors

October, 2016

附录 I 植物中文索引

Appendix I Chinese Species Name Index

附录 II 植物学文索引

Appendix II Scientific Species Name Index